LYTHRUM SALICARIA – GLOBAL INVADER

Factors behind Purple loosestrife's success

Kai Aulio

Grosvenor House
Publishing Limited

This book is published by
Grosvenor House Publishing Ltd
Link House
140 The Broadway, Tolworth, Surrey, KT6 7HT.
www.grosvenorhousepublishing.co.uk

A CIP record for this book
is available from the British Library

ISBN 978-1-83615-372-6
eBook ISBN 978-1-83615-373-3

Contents

Chapter 4

Chapter 5

INTRODUCTION

The never-ending struggle between Good and Evil is as old as humanity, and stories of these opposing forces are culturally mainstream in today's literature, films, and music. Divisions into desired and unwanted objects are also evident in the Plant Kingdom and sometimes even within a single species. The purple loosestrife (*Lythrum salicaria* L.), a plant in the Lythraceae family, is famous for its beautiful and long-lasting flowers. Thus, the plant is a popular garden ornament throughout its native European range. The popularity, unfortunately, turned in the opposite direction when immigrants transported the plant with them when moving to North America in the 1800s. On the New Continent, Beauty soon turned out to be a Beast when it dispersed and colonized territories without natural enemies. And the reign of the invader – and the troubles it causes to biodiversity and the human economy – continues as *L. salicaria* conquers new lands around the globe.

The history of the spread of the purple loosestrife is well-known. Still, the plant's path to becoming an American problem could have been foreseen long before it became a reality. The species received much attention from Charles Darwin, the creator of the evolutionary theory, who discussed the purple loosestrife in many of his books and scientific articles. Decades before any problems with plant introductions were recognized, Darwin described in his revolutionary book, *The Origin of Species*[1], how non-native plants can become a plague, which the purple loosestrife later proved to be.

Darwin described his experiences in Argentina during the famous Beagle ship expedition, noting that "one could showcase domesticated alien plants that have conquered entire islands in less than ten years. Many of the most common plants in the plains of La Plata today, which cover the plains and almost smother all other plants, have been introduced from Europe." This text, published in 1859, is like an ominous prophecy that would come true soon after the first *L. salicaria* had gained a permanent foothold on the East Coast of the United States.

The peculiar floral structure of the plant inspired Darwin to dedicate years of effort and laborious crossbreeding experiments to understanding "the most remarkable fertilization event of any organism."

The rapid conquest of the invasive purple loosestrife has shown some signs of slowing down after the initial decades, but the battle is far from over. Most textbooks, articles, and reports regarding the current distribution of the purple loosestrife are outdated. According to federal government statistics, the alien's spread is still ongoing, with the most recent establishment in Louisiana in 2018, a staggering 187 years after the first sightings.

Thousands of articles and dozens of books have been written about the biology and significance of *L. salicaria*, making the task of compiling an overall picture of the species a laborious one. European enthusiasts find it challenging to access publications by federal and state agencies, research institutes, associations, and societies, where the good and bad qualities of strangers are widely described.

The material used in this book is drawn from articles published in various sources, primarily in scientific journals and reports accessible online. Observations from multiple sectors on the characteristics and effects of the purple loosestrife, which underscore its adaptability and competitiveness, have been selected.

With the help of beetles transferred from Europe, the much-publicized biological control operations have been successful in many places. However, insects do not provide a nationwide weapon against invaders. Leaving natural enemies behind by crossing the Atlantic Ocean is the main reason for *Lythrum salicaria*'s success on the New Continent.

Many more tools are advancing the success of the invader. The purple loosestrife is a powerful plant that competes for space and resources. Little attention has been paid to the growth potential. In North America, *L. salicaria* can grow more than three times taller than it does in Europe; however, its size alone does not explain its success. One of the keys to continental conquest is the massive seed production of the newcomers. Another factor contributing to the overwhelming success of *L. salicaria* is the species' adaptability to local environmental conditions and climate change.

The main reason for suspicions and fear of the alien organism is the impoverishment of natural biota. The indisputable fact is that non-native, colonizing organisms pose a significant threat to natural ecosystems and agriculture worldwide. In North America, up to 90 per cent of threatened species are at risk due to invasive alien organisms. This threat significantly exceeds the much-debated damages caused by anthropogenic alterations in land use or environmental contamination. The myriad of invasive non-natural plants cover 40 million hectares in the USA and expand their reign by 1.2 million hectares yearly.

The scale of the invasion of non-native organisms is alarming. According to the *Intergovernmental Platform on Biodiversity and Ecosystem Services*[2], humans have introduced

37,000 alien organisms worldwide. Of these, more than 3,500 species replace native biota, detrimentally altering the structure of local ecosystems. In natural ecosystems, the invasions of alien organisms are responsible for approximately 60 per cent of plant and animal extinctions.

Besides the irreversible damage to natural entities, the invasion of non-native organisms also causes enormous economic loss to the human economy. Over $ 400 billion is lost worldwide due to the annual triumph of alien invasions. What is most alarming is that human activities directly or indirectly cause the lion's share of this damage.

Considering the enormous losses caused by the invasive organisms, it is understandable that the purple loosestrife has a bad reputation as a rapidly colonizing alien. In the USA, *L. salicaria* is growing in areas covering millions of hectares; however, most of its occurrences are small in scale. Thus, *Lythrum salicaria* could be assumed to threaten only local biota or habitats. In many regions, nature already has the time and capacity to adapt to the occurrence of strangers[3]. Several studies have demonstrated that aliens can benefit local biota – plants and animals.[4,5]

The reality is, however, much direr. The scale of the troubles caused by *Lythrum salicaria* is demonstrated by the costs the plant causes globally. Regarding control and management costs, purple loosestrife is one of the most detrimental invasive aquatic plants worldwide.[6]

The scientific terminology used in this book follows the *Oxford Dictionary of Plant Sciences*[7]. Defining the terms non-native, alien, exotic, non-indigenous, and any other description of an organism that is a "guest" in its present location is a versatile and vital issue. Numerous alternatives are employed, particularly in international literature[8]. Several terms are used in this book, but misinterpretations are unlikely to arise. *Lythrum salicaria,* also known as purple loosestrife, is an unwelcome guest in North America and several regions in other continents.

CHAPTER 1

DARWIN'S NIGHTMARE

In general, nature keeps troublemakers in check

The purple loosestrife (*Lythrum salicaria L.*) is a familiar, widespread, and highly valued wildflower that has also been deliberately propagated by moving and planting abundant and long-flowering plants in gardens. This is the position in Europe, but the species is not considered a beauty on the other side of the Atlantic Ocean. In the New World, the plant is considered – and widely deemed – a monster.

The plant, which migrated and settled in North America alongside humans, has become a despised invasive species on the New Continent, and every effort is being made to eradicate it. However, the newcomer is stronger than the means of human control; therefore, the stranger conquers more territories while destroying the native wetland flora of the continent.

As an explanation for the success of the purple loosestrife, the primary factor is the absence of natural enemies in the New World. In its native range, the plant remains in check due to leaf-eating insects, particularly beetles.

In North America, no group of herbivorous animals has begun to use the newcomer as a food source. Therefore, being free of enemies, the alien has gained a competitive advantage over the region's local flora. Leaf beetles are so effective in curbing *L. salicaria* growth that these insects have been deliberately introduced to the United States for the biological control of the invasive species.

With the absence of plant-eating animals and high seed production, *L. salicaria* has become a textbook model for describing the rapid progression and harm of alien species. The background to these explanations is rooted in the theories and works of Charles Darwin, the foundation of evolutionary biology. As early as 1859, Darwin outlined in *The Origin of Species* how any plant can fill the Earth and displace other species with its established seed production if some external factor – a competing species, a predator, or environmental conditions – does not limit the production of new organisms advancing exponentially.

Jonathan Silvertown, professor of ecology at the Open University, cites *Lythrum salicaria* as a model organism for the unrestricted spread of plants in an enemy-free natural environment. According to Silvertown, the purple loosestrife behaves on the New Continent like a typical *Darwinian Demon*, which can spread uncontrollably, destroying other species found in similar places, even to local extinction. The newcomer has become a nightmare in North America due to its ability to escape its natural enemies in the Old World.[9]

Escaping from herbivores is undoubtedly essential, but by no means the only reason for the extraordinary success of the purple loosestrife on the New Continent. The success of alien species may be based on one or more biological or environmental factors acting simultaneously.

In the summary of published scientific literature on the characteristics of newcomer plants conquering new territories, U.S. scientists Jonathan Fleming and Eric Dibble listed 13 traits that enable aliens to achieve dominance in their new habitats. The extensive review reveals the diverse reasons for the success of alien organisms in conquering new territories.

The purple loosestrife, which has effectively colonized North America, has practically all the means of vegetation available in its arsenal.[10]

Hundreds of articles in scientific journals, newspapers, and magazines have been published about the appearance and distribution of *L. salicaria* in North America. The tone and conclusions have been largely negative, especially in recent decades, following the initial interest in the beautiful plant, which has since subsided. The rapidly spreading alien has gained a questionable reputation as a scourge, even a plague, the transfer and peddling of which is prohibited in many U.S. states. But does the beautiful stranger deserve the discrimination and contempt the plant has received? There is less scientifically valid evidence of the harms caused by purple loosestrife than might be inferred from the species' questionable reputation.

In 1995, Mark Anderson, a researcher at the University of New Hampshire, analyzed 71 published studies on the effects of *Lythrum salicaria* on the native flora and fauna of North America. The result of the summary is surprising: A total of 29 species of organisms were found to utilize *L. salicaria* to their advantage, and not a single plant species was found to have regressed at the national or state level due to the colonization of purple loosestrife. However, hardly anyone disputes the negative influences of the newcomers on the local scale.

For example, the relationships between the alien purple loosestrife and the native broad-leaf cattail (*Typha latifolia*) growing in the same wetland colonies have been studied experimentally. In gardens and small-scale plantings, the purple loosestrife has proven to be a superior competitor to cattails. However, in nature, the ability of the newcomer to displace tall native species has not been observed in the same proportion.

The diversity of flora may even increase in some locations after *L. salicaria* has settled in communities.[11]

A familiar plant with many names

Lythrum salicaria grows in a wide range of habitats. The species is typically found at the water's edge, where it inhabits both salty seawater and inland freshwater, often with its roots and shoot bases submerged below the waterline, as well as in ditches and wetlands. However, in terms of its natural habitats, the species is unpretentious. At the other end of the spectrum, the purple loosestrife can be found high above the waterline in the crevices of cliffs, colonizing inland sand fields or peat soil depressions that remain dry for long periods.

The plant's scientific name, *Lythrum*, is derived from the Greek word lythron, meaning "bloody" or "blood-stained." The species name refers to willows (*Salix*), as the leaf form of the plant resembles the shape typical of willows.

In English, *Lythrum salicaria* is known as purple loosestrife, and what could be more apt than to name the plant after its large flowers, which shine bright purple for months in mid-to-late summer?

Besides its role in natural history and gardening, the plant has also found its place in British literature and art, gaining worldwide fame. In William Shakespeare's *Hamlet* (1599–1601), several plants are listed in the act when Ophelia drowns herself in a river. The plants are mentioned when Gertrude describes the scene of Ophelia's watery demise with the words "long purples that liberal shepherds gave a grosser name/ But our cold maids dead men's fingers call them" (Shakespeare, *Hamlet*: 4.7., 168–171). The name Long Purples was also associated with *Lythrum salicaria* in John Clare's *The Shepherd's Calendar* (1821).

4

It is not guaranteed that the name Long Purples represents *Lythrum salicaria*. However, several analyses consider this plant to be the most prominent candidate. The early purple orchid (*Orchis mascula*) is an alternative species to the long purples. Based on the occurrence and prevalence of these two plants in the riverside meadows, *L. salicaria* appears to be the most probable.

The same conclusion was also made by the artist John Everett Millais, whose painting *Ophelia* (1851–1852) – one of the most admired paintings in the permanent collection of Tate Britain in London – shows erect plants with bright purple spikes lining the river where Ophelia's plant-covered body is resting. According to art historians, Millais made excursions to study typical riverine plant communities when sketching and painting *Ophelia*[12].

In German, the name of *Lythrum salicaria* is Blut Weiderich. The name refers to the flower's colour and the leaf shape, as *Blut* denotes blood, and the remainder of the species name denotes willow-like. In French, the name is *Salicaire commune*; in Spanish, the plant is known as *Lysimaquia roja*.

Immigrants took the Beauty along into the New World

An old-world plant in origin, the purple loosestrife is native to Europe and Asia; however, the plant is now present on all continents except Antarctica. Its range extends from the British Isles in the west to central Russia in the east, reaching the Arctic Circle in the north (in Eurasia, up to 67°N, and, as an alien newcomer, in Canada, up to 51–56°N).

In addition, the species is found in North Africa, Australia, New Zealand, Japan, and the Korean Peninsula, as well as in India and many Southeast Asian countries and parts of the northern Himalayan Mountains. In the Asian parts of Russia, the species' northern limit is at 61°N; in China, the range extends to 50°N. The southern limit of *Lythrum salicaria* in Eurasia is 24°N in China and 33°N in Afghanistan and Iran. In Australia, the range spans between 23°S and 42°S and 137°E and 153°E[13].

The Old-World herb has already reached South America as well. In 2015, the first spontaneous population of the purple loosestrife on the continent was reported in Chubut Province, Argentina[14].

The purple loosestrife's latest conquest was South Africa, where the alien has established seven small-scale bridgeheads. The future of *L. salicaria* in South Africa seems bright. Success in reproduction is the key to spreading and colonizing new territories. The complex structure of the weed's flower might be a limiting factor in finding suitable pollinators. But the alien overcomes the obstacles. The study found four native and one alien taxa – the Cape honeybee and African Monarch butterfly being the most important – successfully fertilizing *Lythrum* flowers.

In addition, the seed production of the purple loosestrife was successful even without pollinator insects. Self-compatibility resulted in viable seeds, although cross-fertilization by pollinator insects produced a significantly larger number of seeds. Local pollinators adapt well to the polymorphism in *Lythrum* flowers. All three style types are present in South African *Lythrum* populations, producing "huge amounts" of viable seeds. Based on its reproductive capacity, the authors conclude that "this species has the potential to become a significant invader of rivers and wetlands in South Africa."[15]

New Zealand is one of the target regions where *Lythrum salicaria* has been established as an alien species. The non-native weed has been recorded in over two hundred locations throughout the island nation. The appearance and spread of

L. salicaria are taken seriously. The risks that the aggressive non-native plant pose to natural biota are so high that every person who sees or suspects the presence of the plant must immediately report the sighting to local authorities.

The newcomer is unwanted, and the purple loosestrife is listed in the *National Pest Plant Accord Species*[16] and the *List of Environmental Weeds in New Zealand 2024.*[17]

The purple loosestrife has received the most publicity as an unwanted newcomer in North America. In the first half of the 1800s, European colonists introduced *Lythrum salicaria* to the New World – deliberately as seeds of an ornamental plant and unintentionally accompanied by the solid ballast masses of ships. The first, non-specific, mention in the literature was in the *Flora Americae Septentrionalis* in 1814 (ref. Stuckey 1980[18]).

It was not until the 1980s that the U.S. Fish and Wildlife Service investigated the extent and significance of the newcomer's areas of occurrence, and the stranger was declared an outlaw. Classified as a severe non-native species, the purple loosestrife nowadays grows in all continental U.S. states except Florida and in at least ten Canadian provinces or territories. In the United States, *L. salicaria* covers an estimated two million hectares, and the species occupies an average of 190,000 hectares of wetland area annually.[19]

Besides being exceptionally rapid, the distribution of *L. salicaria* in North America extended farther than most alien immigrants. In the history of non-native plants introduced to the U.S. via ballast masses on the continent's eastern coast, most newcomers did not spread far from the coastline. Only a few even survived for long times in their new ranges.[20]

The story of the alien continues as the conquest proceeds, and so do the updates and warnings of the noxious invader.[21]

Due to the damage caused by *Lythrum salicaria,* the invasive species has been nicknamed the *Purple Plague* in the United States. The purple loosestrife is included on the list of the world's 100 most dangerous alien species by the International Union for Conservation of Nature (IUCN).[22]

The newcomer's impact on North American nature and economy is the subject of lively debate in the media, social media columns, and scientific publications. Comparisons of several studies have revealed that *L. salicaria* is both a friend and a foe in North America, i.e. the alien has detrimental effects on some locations. Still, in other destinations, the invader is well-suited or positively influences the natural diversity. Today, asking if we should be concerned about the purple loosestrife is hardly necessary.[23]

From beautification to a hated intruder

In Europe, *Lythrum salicaria* is valued and admired as a beautiful plant; however, the situation is quite the opposite in North America, especially in the United States. The competitive advantage of the purple loosestrife – a plant intentionally transferred to the New World from Europe – unexpectedly became an enemy to the natural habitats in the resident areas. Transferring or selling the plant or its seeds is now illegal in at least twenty states.

European insects, which use purple loosestrife as a food source, were introduced to the United States to combat the invasive species. Nowadays, at least 45 million dollars is spent annually in the United States on controlling aliens.

Despite the widespread publicity and occasional dramatic headlines of the damage *L. salicaria* has caused, the problems are not severe nationally. More than 50,000 species of organisms are classified as aliens in the United States, many of which cause ecological and economic issues. The tens of millions of dollars spent annually on mechanical, chemical, and biological control of *L. salicaria* seem small compared to the $1.5 billion spent annually on yard lawns, parks, and golf courses to control dandelions and other common weeds.[24]

The extensive control tasks to combat *Lythrum salicaria* have been criticized, as only a few scientifically valid studies have been conducted to describe the newcomer's harmful effects in the early stages of management. The poor results of pesticide treatments (poisonings) initiated in the mid-1980s were not published, and much of the activity was based only on assumptions. However, ecological and economic studies have now confirmed the harmfulness of the alien plant. Today, the benefits of various control measures significantly outweigh the possible local harms.[25]

A stranger grows into a giant

In Europe, the shoot length of purple loosestrife seldom exceeds 1.5 metres, but in North America, *Lythrum salicaria* can grow significantly taller. In American botanical descriptions, the height of the purple loosestrife is reported to be up to 10 feet, approximately 3 metres, with a maximum height of 3.5 metres.[26]

The enhanced size of *Lythrum salicaria* on the New Continent is at least partly due to the new gene combinations resulting from the crossbreeding of genomes from distant origins, a phenomenon known as *heterosis*.

Due to the considerable shoot length and high crop density, the production and above-ground biomass of *L. salicaria* populations are exceptionally high. For example, in the wetlands of the Delaware River watershed, the maximum shoot biomass of purple loosestrife at the end of the growing season exceeded 2.1 kilograms of dry matter per square metre.[27]

Chapter 2

Spectacular plant with infinite qualities

"The most significant reproductive event of any organism"

The purple loosestrife has three kinds of flowers, with varying lengths of pistils and stamens, as well as a wide range of flowering times. For successful fertilization, pollinator insects must locate other plant individuals of the same size and with a similar ratio of pistils to stamens. Between the numerous flowers of any individual, pollination will not succeed.

Lythrum salicaria ensures continuous cross-pollination, essential for its vitality, with the help of reproductive organs of varying sizes and opening times. The secrets of the complex flower structure and reproductive biology of the purple loosestrife were discovered by Charles Darwin in 1864. In an article published in the *Botanical Journal of the Linnean Society*, he praised the three types of flowers of the *Lythrum* species as providing "perhaps a more substantial fertilization event than any other plant or animal."

Besides detailed descriptions of the floral structures, Darwin understood the importance of size differences in stamens and pistils, as well as the varying opening times of the stamens, in securing cross-pollination.[28]

After his morphological observations, Darwin conducted hundreds of crossing tests with various stamen–pistil combinations, ensuring the functionality of cross-pollination as explained by the theory. The variation in the size and position of stamens and pistils ensures that pollen adheres in precisely the correct position in the fertilizing insect's body to catch the female flower's stigma.

The strategy that ends up in cross-pollination works best for flowers with a short style of the female carpel, located deep at the base of the horn-like corolla. When the pollen-receiving pistil stigmas are located as deep as possible in the corolla horn, as in the purple loosestrife, the insect always approaches the pistil in the correct position.[29]

Darwin's conclusions about the importance of the different positions of the stamens and pistils have since been experimentally confirmed. The most thorough experiments on the different flower forms of *Lythrum salicaria* and the success of cross-pollination were conducted by a team of researchers led by Joana Costa, a researcher at the University of Coimbra in Portugal. They also pointed to several misconceptions leading to the failure of previous crossbreeding experiments. Despite the differences in size and location of the reproductive organs and varying opening times, self-pollination is still possible in purple loosestrife. However, it is less common than cross-pollination.[30]

Cross-pollination should be ensured

In its complex and functionally appropriate reproductive organs, *Lythrum salicaria* has three types of flowers: (1) flowers in which the body of the pistil is longer than the 12 stamens in two whorls (6 + 6), (2) flowers in which both long and medium-length stamens are longer than the style of the pistil, and (3) in the third type, the filaments of some stamens are shorter than pistil's body, and the stamens of the second whorl extend higher than pistil's body.

Besides the length of the filament, the colour of the stamens is also variable. Long stamens are green or greenish, medium are dirty yellowish, and short are yellow. Ensuring cross-pollination is "almost incredibly complex" for purple loosestrife, as the pollen grains produced by stamen types vary in size and colour, and the reserve food they contain, which insects desire, also comes in different forms.

Pollen grains germinate and fertilize the pistil only if transported and received by a flower type different from the stamen from which they originate. This procedure ensures cross-pollination. Since each individual of *Lythrum salicaria* has only one kind of flower, fertilization and seed production should only occur in another plant individual.[31]

Self-pollination is possible

Despite complex precautions, some *Lythrum salicaria* also produce seeds through self-pollination. Experiments have demonstrated that the pollen produced by a flower's stamens can fertilize the pistil of the same flower.

Researchers Christopher Balogh and Spencer Barrett from the University of Toronto in Canada demonstrated in a two-year experiment that pollen, especially from medium-sized stamens, can end up in the stigma of the same flower's pistil. This means that the plant individual pollinates itself.

Approximately 34 per cent of *L. salicaria* in the research data produced seeds through self-pollination, although cross-pollination is the predominant method of fertilization. Of the three types of flowers, those with medium-sized stamens produced more seeds than the others through self-pollination.

The offspring of individuals that developed through self-pollination retained the characteristics of the parent plant, i.e., more self-pollination than average was observed in the next generation as well.

The humidity of the growing site (dry vs. wet) did not affect the occurrence of self-pollination. Maintaining a stable self-pollination indicates that the phenomenon is hereditarily determined. Between plant individuals from different populations, on the other hand, there were significant variations. This further reinforces the genetic background of the self-pollination.[32]

Newcomer steals pollinators

Alien species that have spread to new areas may also indirectly alter the relationships between native plants, attracting flower-pollinating animal visitors. The purple loosestrife is often accused of stealing pollinators in North America, where *Lythrum salicaria* frequently shares habitats with the region's native species, the winged loosestrife (*Lythrum alatum*).

A team led by Ohio's Kent State University researcher, Beverly Brown, investigated pollinator insect behaviour and flower preferences in an experiment where both *Lythrum* species were grown as single-species stands and as mixed communities of both species. The density of plant individuals (number per unit area) varied in both cases.

Increasing the density of *L. alatum* shoots in the plot decreased the number of pollinator visits per flower. However, at the end of the growing season, the number of seeds produced per flower had remained the same. In conclusion, there were sufficient pollinators in all circumstances to ensure maximum reproductive input, regardless of the number of plants.

When the two species of *Lythrum* were grown as mixed crops, the flowers of *L. alatum* had significantly fewer pollinators than the pure, single-species population of the species, and there was also an apparent reduction in the number of seeds developing per flower. The decreased seed output resulted from the competition that *L. alatum* lost against the more attractive flowers of the alien intruders.[33]

Alien weakens native flora

The competitiveness of *Lythrum salicaria* in new, suitable habitats is considered a severe threat to local plants, but the effects in nature are mixed. Madlen Denoth and Judith Myers, researchers at the University of British Columbia in Canada, studied the success of the local plant species, Henderson's checker-mallow (*Sidalcea hendersonii*), which was found only in this region after the purple loosestrife had begun to conquer the wetlands inhabited by the local species. *Sidalcea* is a species in the mallow family (Malvaceae), characterized by dense inflorescences and red flowers resembling purple loosestrife.

Monitoring every two years for 20 years revealed that the number of individuals in the native flora had declined to approximately half of the baseline. At the same time, the number of alien *Lythrum salicaria* increased by 20%. Visible deterioration of the shoots mirrors the weakened condition and decline of *Sidalcea*. Still, flowering and seed production, vital for the species' long-term survival and spread, were not affected by the conquest of *Lythrum salicaria*.

The researchers further investigated the mechanisms and effects of competition between the purple loosestrife and the endemic species in experimental conditions by removing *Lythrum salicaria* and other local plant species near the endemic species. Removing neighbouring individuals gives the plant space to grow and improves lighting, water, and nutrient conditions. Thus, reducing the plant cover will help the taxa, which was the management's target. However, the removed species still retained a vital position in the habitat.

Endemic *Sidalcea* also improved after *Lythrum salicaria* and several native species were removed from the area. Researchers concluded that, despite the local competitive situation, the purple loosestrife is not a threat to the survival of the endemic *Sidalcea hendersonii* in the wetlands of British Columbia.[34]

A rich seed bank does not guarantee an unlimited spread

Up to 2.7 million small-sized seeds develop from one individual shoot of *Lythrum salicaria* each year. The seeds are between 200 and 400 micrometres (0.0002–0.0004 millimetres) in size and weigh 0.5 to 0.6 milligrams, meaning that one gram can hold 500 to 600 seeds.[35]

The seeds of the purple loosestrife do not float in the water, but the currents carry the reproductive bodies long distances. The seeds spread mainly via air currents, and in winter, also on snow- and ice-covered surfaces.

An efficient means of *L. salicaria*'s long-distance dispersal is seed transport by birds. A study in the Netherlands showed that waterfowl can transport the small-sized seeds of the purple loosestrife for hundreds or even thousands of kilometres.[36]

The vitality of enormous seed production is increased because the seeds remain capable of germinating in the soil for several decades. At least 80 per cent of the seeds can germinate for 2–3 years. Soil characteristics do not matter, as the seeds of the purple loosestrife survive even in highly acidic (pH 4) to very alkaline (pH 9.1) conditions. A seedling growing from one seed can develop 30 to 50 shoots. The hundreds of flowers in each spike inflorescence open in phases over several months, so the pollen and nectar season, attracting pollinating insects, is prolonged.

In addition to vast seed production, a plant that conquers new territories requires suitable conditions for germinating its seeds. Each plant taxon has specific and highly variable requirements. One would think that millions of purple loosestrife seeds could be found everywhere near already-established flora. In a U.S. study, the topsoil of the wetland vegetation had an average of 410,000 purple loosestrife seeds. However, the seeds remain mostly at their point of origin, despite their potential to spread. Most *L. salicaria* seeds settle in the soil within a radius of less than three metres (ten feet) from the parent plant.[37]

Despite the enormous seed production, the capacity for sexual reproduction of the purple loosestrife can be unexpectedly low. Researchers from Queen's University in Canada surveyed the number of purple loosestrife seedlings in ten wetlands where the alien invader was present. They compared the number of seedlings of plants in 11 "*Lythrum*-free" wetlands.

Lythrum salicaria seeds were not found in many areas where the species had not settled. Despite the seemingly limitless amount of seed production, it turned out that there are restrictions on spreading the purple loosestrife. On the other hand, the species' seeds were found and developed into seedlings on all plant-free soil patches in those riparian colonies where *L. salicaria* was already well established.

Soil conditions and other environmental parameters were more critical in determining seedling development than the spread of *L. salicaria* seeds. Scientists interpreted the development of new seedlings as being more influenced by soil and other environmental conditions than by the spread of *Lythrum salicaria* seeds. The importance of ecological conditions was emphasized by the observation that the seeds of the broadleaf cattail (*Typha latifolia*) developed into seedlings. Still, the germinated seeds of the purple loosestrife did not develop into seedlings.

Canadian researchers concluded that the differences in plant development are due to species-specific growth characteristics rather than interspecific competition. The competition between wetland plant seedlings is constant and intense, but the struggle for space and resources is equally stringent for all species.

The competition for space between seedlings and established plants later favours the purple loosestrife solidly. Plant diversity is lower in wetlands dominated by *Lythrum salicaria* than in meadows where the purple loosestrife is not established.

The Canadian research concludes that preventing seed production is the most effective solution to avoid the continuous and unwanted spread of the purple loosestrife in North America.[38]

Time of propagation is a strategic choice

Along with the time of flowering and seed production, seed germination influences the purple loosestrife's ability to propagate. According to American sources, the seed germination rate ranges from 50% to 100%.

In a follow-up of two consecutive growing seasons, North American *Lythrum salicaria* seeds remained attached to the parent plant for a long time at the end of the growing season. The seeds are released and spread over 6 to 8 weeks when the average air temperature has dropped below 15 degrees Celsius (59 degrees Fahrenheit).

Seed germination is significantly enhanced after being in a cold and humid environment. Germination occurs only at the beginning of the next growing season or often years after a period of rest in the soil as a seed bank.[39]

Least germination occurs in seeds that mature early and have not experienced cold conditions. Seeds that have matured at the end of the growing season, remained attached to the parent plant, and experienced cold exposure before shedding may germinate during the same growing season. Keeping mature seeds attached to the parent plant is a beneficial strategy for the purple loosestrife, ensuring stepwise germination and thus taking advantage of the best climatic and environmental conditions.[40]

Seedling trade damages nature

The status of *Lythrum salicaria* as a noxious weed has led to its outlawing in many U.S. states. Transplanting purple loosestrife into gardens and selling seedlings or seeds of the species are prohibited in at least twenty states. Despite the prohibitions and recommendations, the beautiful-flowered plant remains a popular choice as an ornamental plant, and garden stores and wholesalers meet the demand with an abundant supply, despite the prohibitions. Often, purple loosestrife gets unintentionally introduced into gardens in plant mixtures.

University of Minnesota researchers Kristine Maki and Susan Galatowitsch investigated the availability of purple loosestrife and, above all, the prevalence of non-native species that inadvertently end up in gardens. The researchers placed 44 flower and seed orders from 34 established horticultural companies across the U.S. to determine how often and how many banned plants were delivered to a given address in Minnesota. Illegal or restricted plants came in almost all orders.

As many as 92 per cent of shipments included unwanted and prohibited plants or plant parts, along with the desired orders. *Lythrum salicaria* was a common extra to the purchases. Among the plants classified as harmful or prohibited wetland plants in Minnesota, the shipments included, in addition to purple loosestrife, the flowering rush (*Butomus umbellatus*), the European frogbit *(Hydrocharis morsus-ranae)* and the curled pondweed (*Potamogeton crispus*).

Ninety-three per cent of the shipments from garden stores included plants that were not ordered. One in ten shipments contained plants whose sale and delivery to Minnesota are prohibited by law. The most common unwanted and restricted species was the common duckweed (*Lemna minor*), a small, free-floating plant.

Plants unidentified by even the expert botanists were found in almost one in five orders, while unordered seeds were found in 43 per cent of shipments. The Minnesota researchers conclude that the trade of banned species is a serious driver of the spread of harmful aquatic and wetland plants in the United States.[41]

"Beautiful rascal" – embarrassing but lovable

Despite prohibitions, the trade in the beautiful purple loosestrife is brisk in the United States. With the popularity of the spectacular and easy-to-care-for flowering plant, variants similar to their wild ancestors have been developed. These varieties are also prohibited. Due to its popularity, not only are garden enthusiasts interested in the purple loosestrife but so are seed and seedling merchants.

The purple loosestrife is marketed widely and spectacularly to various destinations. Unfortunately, the ecological dangers posed by strangers have not been sufficiently identified, or at least, not recognized.

As recently as the early 1980s, horticultural experts and enthusiasts praised the plant, continuing to plant it in their gardens – albeit acknowledging the newcomer's potential risks as a weed – with colourful expressions like "beautiful rascal." In other words, the newcomer was known to cause problems, but such minor concerns should not be overlooked.[42]

Improved varieties interbreed and spread into nature

Varieties of purple loosestrife bred as garden plants have been developed with the intention that these ornamental plants are sterile. The abundantly flowering and vigorous garden plants typically produce pollen, fertilizing both bred and wild-growing *Lythrum salicaria.*

Believed to be incapable of reproducing and therefore harmless, plants of purple loosestrife born through crossbreeding have accelerated the colonization of the alien species in North American nature.[43]

Chapter 3

A superior competitor

The size makes the difference!

In determining interspecies relations, it is essential to distinguish competition between individuals on the ground (shoot race) and under the ground (root system race). In terrestrial competition, fast-growing species are particularly successful as their shoots develop taller than those of other plants, are denser in terms of the number of individuals, and have a greater shading capacity. In underground competition, the growth rate and the number of horizontally growing rhizomes determine which plants can occupy the habitat.

The role of competition between *Lythrum salicaria* and other plant taxa was studied in Canada by researchers from the University of Ottawa. The field study examined the above-ground and underground competition between individuals of *L. salicaria* and the fringed sedge (*Carex crinita*) over several years on a gradient with varying environmental conditions in Ontario's Ottawa River basin. The extremes of the comparison were an open sandy beach that maintains few plant individuals and a sheltered bay dominated by an extensive plant cover. Shoot (above-ground) and root (underground) biomasses were measured from the plants in 60 plots along a line transect.

When analyzing the biomass of sedges and purple loosestrife in microhabitats with varying environmental conditions, interspecific competition intensified directly in relation to the amount of plant biomass. Aboveground biomass production was the most critical variant in defining each species' competitive strength. In this respect, *L. salicaria* outnumbered most of its competitors.[44]

Remarkable weakening of the competitors

Regarding a plant's competitiveness, it is worth exploring (a) the possibility of suppressing or displacing other species and (b) resistance to pressure from other species. In this assessment, purple loosestrife has proven to be a superior performer.

In an experimental study conducted by the team at Ottawa University, seven wetland plants were first grown in monocultures for three years. In the second phase, 48 species were planted within each of the seven monocultures, and the survival of the transplanted plants was assessed in the test plots over a four-month period. As a control, each test plant was grown alone, without competition with other taxa. At the end of the experiment, the height and shoot biomass of each transplanted plant were measured in test plantations and control plots.

As expected, the vital monocultures effectively suffocated transplanted individuals of other plants. The adverse effects in all dense monocultures reduced the growth and production of transplanted species to less than half of the optimal development that each species would achieve in the absence of competition. In dense and established plantations, the purple loosestrife caused the most growth-reducing effects compared to 48 wetland plant taxa.

Lythrum salicaria reduced the growth and dwarfed others more than any other plant studied, even more significantly than emergent plants that grow markedly taller than the purple loosestrife. At most, the purple loosestrife reduced the production of other wetland plants by more than 80%.[45]

In the United States, the ability of *Lythrum salicaria* to suppress the establishment of other plant species was experimentally studied. The tests included a collection of 39 typical wetland and sedge meadow plants with diverse life forms. Of the *Lythrum*'s competitors, 75 per cent were native species commonly found in the usual habitats of purple loosestrife. Multiple degrees of soil fertility and damage caused by artificial herbivory (plant-eating animals) or chemical control treatments were included. The main target of the loosestrife's competition was the broadleaf cattail (*Typha latifolia*), a native, tall, emergent wetland plant. *L. salicaria* and *T. latifolia* were grown in monocultures in standardized microcosms, and the possibilities of the rivals to establish in these stands were analyzed.

The purple loosestrife was a superior competitor in every aspect of the plant's growth characteristics, including a 50 per cent reduction in the competitors' biomass and a 34 per cent reduction in the species diversity of other plants in the experimental plots with *L. salicaria*. The main result was that the purple loosestrife suppressed plant communities much more effectively than the broadleaf cattail, despite the latter being a markedly taller and more robust plant than the alien loosestrife.

This was the first study to demonstrate that an invasive, non-native plant can suppress colonizers more effectively over time than a dominant, local native species.[46]

In Canada, the competitive superiority of *L. salicaria,* especially in nutrient-rich wetlands, was highlighted in the experimental study. In the phytometer study of 26 shoreline plant species, the purple loosestrife was the most potent eliminator of a reference plant, the ditch stonecrop (*Penthorum sedoides* L.). The two-year experiment revealed a consistent pattern in the competitive performance of the highly variable mixture of shoreline plants. The best performers were the large, leafy plants typical of fertile habitats.

Under high nutrient supply and after one growing season, the purple loosestrife caused a 96% reduction in the biomass of *Penthorum*. When grown in nutrient-poor sandy soils – an atypical rooting medium for *L. salicaria* – the purple loosestrife's power to wipe out the competitive plant was near the average of the study material.

The species-to-species relationships changed during the second year. After two growing seasons under high nutrient conditions, seven tall emergent macrophytes exceeded *L. salicaria* in reducing the biomass of *Penthorum sedoides*. At the top of the list was *Typha glauca*, which choked the ditch stonecrop by 99%. Competition with purple loosestrife weakened the competitor's growth by 89%. However, contrary to the average performance in the first year under low-nutrient conditions, *L. salicaria* was now the strongest competitor, reducing the biomass of *Penthorum* by 99 %.[47]

Nature does not follow experimental results

The colonization and establishment of *Lythrum salicaria* rarely result in uniform monocultures in wetland habitats. Still, the intense competitor is considered harmful to other plants. Several controlled experimental studies have validated purple loosestrife's ability to suppress the growth of other wetland plants. However, in nature, the competitive situation may be completely different.

Heather Hager and Rolf Vinebrooke, researchers at the University of Regina in Canada, analyzed the ratios of wetland plants at ten habitats in Minnesota, USA. The study compared the diversity of plant species in sites dominated by the purple loosestrife and in areas dominated by cattails (*Typha* spp.).

Growing in mixed communities, the purple loosestrife was undeniably more vigorous than other species and reached dominance. However, in wetlands with varying environmental conditions, *Lythrum salicaria* was found to colonize several types of microhabitats. Due to its ability to withstand various environmental conditions, the species can colonize multiple habitats. The surprising conclusion of the study was that vegetation diversity might be at its highest, contrary to most prevailing views, in communities where the purple loosestrife is present.[48]

Similar results, showing that *Lythrum salicaria* outcompetes native flora, were confirmed in a literature review by the University of New Hampshire researcher Mark Anderson. According to analyses, the colonization by the non-native *Lythrum* species did not harm the plants that initially occupied the habitats. However, the newcomer can locally suppress native flora or other aliens, forming dense monocultures.[49]

The reputation of *Lythrum salicaria* as a noxious alien that can wipe out native plant communities appears one-sided and may be valid only in specific environments. Experimental studies by Janet Morrison, a researcher at the College of New Jersey, demonstrated that alien and native herbs coexist in healthy wetland communities. Diversity – i.e., the number of plant taxa per unit area – remained high even if the coverage of *Lythrum salicaria* increased up to half of the surface area.

Removing the non-native loosestrifes from experimental quadrats did not restore dominance to native plants. In contrast, excluding *L. salicaria* led, in some cases, to the dominance of another non-native species, the gramineous reed canary grass (*Phalaris arundinacea*).[50]

Dominant alien was replaced in 10–15 years

Freshwater wetlands – the most favourable habitat for the purple loosestrife – are sensitive to disturbances caused by the colonization of new taxa. *Lythrum salicaria* is one of the most competitive invaders in these ecosystems, but species-specific superiority is only one factor in determining the outcome in nature. While studying 16 temperate wetlands in the US, researchers at Pennsylvania State University, Tara Mazurczyk and Robert Brooks, demonstrated, against the prevailing "textbook wisdom", that the dominance of an alien invader can increase nature's biodiversity.

The colonizer can be a strong competitor and gain dominance rapidly, but the reign of this new ruler can be temporary. The dominant species is often wiped out by another, typically an invader or a native species of the natural biota, within 10 to 15 years.

The change in species composition is primarily determined by the natural background quality rather than the characteristics of the species. Against the current scientific dogma, an alien species can be a secondary pioneer in succession.[51]

Growth intensifies in a new environment

For over 150 years in North America, *Lythrum salicaria* has established itself as a strong competitor on the continent. Especially in nutrient-rich and moist habitats, the American populations of the species differ significantly from their European counterparts. Researchers from the University of Wisconsin in the United States and the Czech Academy of Sciences analyzed native Central European populations of *L. salicaria* and communities of the species established in America. The comparison showed that the newcomer populations were stronger competitors than the plants in their native area.

According to the hypothesis predicting the development of competitiveness, the purple loosestrife that has settled in America would benefit from the absence of natural enemies on the New Continent. Thus, the assumption was that the alien plants would grow taller ramets, denser populations, produce greater biomass, and experience more intense flowering and seed production – altogether, forming more vigorous communities in the new grounds than in their places of origin.

Theoretically, *L. salicaria* that have settled as non-native aliens in America should have taller shoot sizes and greater crop density, plant production, biomass, fertility, and seed production than their European counterparts.

In controlled analyses, many predictions were proved to be valid. When growing in physically and climatically similar habitats, the densities of purple loosestrife shoots are higher in American populations than in European populations. On the other hand, the heights and weights of plants are generally identical for most habitat types, especially when growing on moist substrates. In nutrient-rich microhabitats with minor species-to-species competition, the purple loosestrife can grow twice as tall in America as in Europe.

In addition, the American populations of *L. salicaria* are more fertile than their European counterparts. In typical moist shoreline habitats on the New Continent, the proportion of flowering ramets (shoots) is higher in the newcomers' populations than in the species' European relatives. No differences in fertility were recognized when growing on nutrient-poor, not ideal for the purple loosestrife, sandy soils.

Seed production, essential for the species' ability to spread, is significantly higher in alien populations than in European communities of the same species. Especially in nutrient-rich wetlands – the most typical habitats of the purple loosestrife –

the species' seed production in North American populations exceeds that of the species' native grounds. In the New World, *Lythrum salicaria* produces an average of 89 to 103 seeds in each fruit, compared to 58 to 64 seeds per fruit in European populations.[52]

The importance of abundant seed production is further emphasized in comparison to the quantities of seeds that mature and are spread into nature. In European populations, the fruits and seeds of *Lythrum salicaria* are eaten by many plant-eating animals (herbivores), but in North American populations, the purple loosestrife has no natural enemies.

The role of habitat quality, particularly soil nutrient status, on the growth potential of purple loosestrife was investigated in a study conducted in Korea and the United States. In the experiments, the shoots grew more slowly in the alien populations, but the plants grew taller than those from native populations. The morphology of the newcomer shoots differed from that of the individuals in the native populations. Along with the taller shoots, more branches developed in "alien" *Lythrum* shoots, and their leaves were more extensive than in the native populations.

In addition to seed production, vegetative (asexual) reproduction – through roots and rhizomes, detached parts of shoots, or germ buds – is essential in conquering new grounds. Under experimental conditions, vegetative reproduction and propagation of *Lythrum salicaria* were more effective in non-native populations than in individuals from the native population.[53]

Even small populations are stable

The unique structure of the purple loosestrife flowers, formed by stamens and pistils of different lengths, is rare in nature. The structure's purpose is to prevent self-pollination, which requires the system's stability even in changing conditions. Due to cross-pollination, plants constantly receive new genetic material, and gradual changes are inevitable. A widely held assumption is that the stability and survival of the complex structure of three stamens with varying lengths would require a sufficiently large population.

The persistence of the three-stamen flower structure was studied in Finland in the 1970s. Researchers Olli Halkka and Liisa Halkka at the University of Helsinki analyzed the conformity of the flower structures in separate *Lythrum salicaria* populations across 16 islands. The study showed that the three-stamen design was similarly repeated on every island. The conclusion was that gene exchange – i.e., pollination between populations – had occurred between the islands despite large distances.[54]

In his model experiments, Ivar Heuch, a Norwegian researcher in applied mathematics at the University of Bergen, concluded that a three-fruit flower structure could remain stable even if the number of individuals in a population were as low as 20 plants. If the population shrinks to a smaller size than this, the morphology of the flower will slowly begin to change. The first to disappear are the shortest stamens, the only type with a unique gene regulating this phenomenon.[55]

Flowering starts earlier in the El Niño years

The species' remarkable adaptability to sudden and unpredictable environmental changes dramatically enhances its chances of conquering new habitats. El Niño weather phenomena, which regulate the climatic conditions of the entire planet and cause significant changes, are recurrent, but their incidence is irregular.

The seven-year cycle is considered the typical schedule of the weather phenomenon, but there is considerable variation in the timing of events. *Lythrum salicaria*, with its genetic adaptation to unexpected and abnormal climatic and weather conditions, demonstrates a resilience that commands respect.

The growth schedule and flowering success of purple loosestrife were studied in Canada during the intense El Niño year of 1998. The results could be compared to a similar study conducted in the North Bay wetlands near Toronto during the 1997 growth period, which can be considered a "normal year" study.

The 1998 El Niño-Southern Oscillation (ENSO) weather phenomenon was particularly intense, with average temperatures that year higher than usual and rainfall below average. The purple loosestrife's reactions to these changes, which significantly determine the growing conditions, were noticeable. The unusual conditions appeared to be favourable for the plant.

The exceptionally warm spring of 1998 caused the purple loosestrife to bloom on average 14 days earlier than in the typical year of 1997. A similar change was observed in the production of vegetative shoots. In the El Niño year, the production of purple loosestrife biomass was more strongly weighted than usual at the beginning of the growing season. Nevertheless, early spring did not significantly impact the summer's total production.

The total biomass of the shoot (above-ground parts) produced during the growing season, the length of the spike inflorescences, the biomass of the inflorescences, and the autumnal wilting coincided with the average and El Niño years. A change in the timetable of a plant's development cycle in exceptional circumstances can significantly impact the success of the purple loosestrife if the growth and flowering period of the alien is more successful than that of native plant species competing for the same resources.[56]

Success is due to the rapid evolution of the life cycle

The purple loosestrife, a European colonizer introduced to North America, has become a serious nuisance on the continent, conquering new territories and displacing the native coastal and wetland flora at an alarming pace. Over the decades, efforts have been made to control the spread of the newcomer mechanically by weeding, poisoning, and repatriating European insect pests to prevent the growth of the purple loosestrife. Various laborious and costly control measures have produced only local success, and the purple loosestrife has continued to colonize new territories.

The newcomer's success has been attributed to the absence of animals that eat the plant. However, the plant has merits due to its exceptional distribution across the continent. An important factor is the evolutionary development, accelerating the life cycle of the purple loosestrife. The flowering schedule of *Lythrum salicaria* has undergone significant changes in the past 50 years. This is a remarkably brief period in terms of evolutionary development.

Robert Colautti and Spencer Barrett, researchers at the University of Toronto's Department of Ecology and Evolutionary Biology in Canada, presented in the journal *Science* how the life cycle of *L. salicaria* – especially the timing of flowering – has adapted to suit local conditions. The researchers compared the success of purple loosestrife through direct field observations and experimental studies on the east coast of North America, spanning a zone approximately 1,000 kilometres in length, from south to north.

For experimental analyses, seeds of the purple loosestrife were collected from three zones: (a) the southernmost populations of *L. salicaria*'s North American range, (b) the northernmost occurrences of the species, and (c) the populations found halfway between them.

The study meticulously tracked the development schedule of plants from three distinct populations, from germination to the onset of flowering and the subsequent seed production. It revealed significant differences in local adaptation between northern and southern *L. salicaria,* underscoring the species' potential for success and future spread in North America.

Plants derived from northern seeds and grown under northern environmental conditions adapted well and developed to the flowering stage approximately three weeks faster than those grown under the most southerly physical conditions. The accelerated growing cycle benefits the seed production of the purple loosestrife. Early flowering is advantageous in the North, where the annual growing season is short. Another remarkable advantage of accelerated growth is that the number of insect pollinators is at its maximum at the plant's flowering time. Thanks to its rapid life cycle, the purple loosestrife produces up to 20 to 30 times more seeds in its northernmost habitats than in more southern conditions, where the growth rate is slower.

Conversely, *Lythrum* plants grown from southern seeds under southern environmental conditions exhibited a markedly slower development, a strategy well-suited to the longer growing season. This extended period allows individuals to grow significantly taller than in northern conditions, and the vegetative rhizome propagation becomes more effective. The purple loosestrife's ability to adapt its growth strategy to the length of the growing season enables it to colonize new habitats, even in harsh conditions.[57]

Adaptation is crucial for success

Interspecies differences caused by climate change were investigated by comparing the adaptation reactions to changes in humidity and temperature in *Lythrum salicaria* and a member of the North American natural flora, the Drummond rockcress (also known as Canada rockcress), *Boechera stricta*. Several long-term studies of the phenological growth cycles of these two species were used as data for the model study.

Based on published research results and the model, climate change is expected to cause significant changes in the flowering schedules of seed plants. The timing of phenology is a crucial factor in the adaptation of plants to their annual life cycle. In this regard, the two species are markedly different from each other.

Boeschera's flowering starts earlier as temperatures rise. This can be expected to decrease the viability of populations, even if the plant can adapt to the new conditions. The ability to adapt is genetically determined. On the contrary, the beginning of flowering is delayed in *Lythrum salicaria*. Due to an extended growing season, the long-term competitiveness and vitality of plant populations can be expected to improve with climate change.

According to Canadian researchers, the appropriate adaptation to local conditions of the growth and flowering schedules is a crucial factor in the success of purple loosestrife. This genetic characteristic is even more important because the species does not face the same herbivorous enemies in North America as in its native European regions.

Understanding the life cycle strategies, specific adaptation mechanisms, and the speed of development of plants is crucial in the face of rapid and extensive environmental changes. Climate change has significantly impacted the spread and colonization of alien organisms, creating variable growth conditions and increased competition for wild and alien species introduced by humans.[58]

Aliens tolerate variability

An organism spreading to new areas must adapt to the local climate, soil, water, and nutrient conditions, as well as competition from the native species inhabiting the site. The greater the differences between the place of origin and the new habitat, the more flexible biological adjustment mechanisms are required from the newcomer. Once established, the colonizer typically expands its habitat and range, which may require new adaptability.

The adaptation requirements are particularly evident when alien species spread along the south-north direction, as both climatic conditions and the length of the day will fundamentally change. The gradual change in living conditions necessitates that non-native species adapt to the new schedules related to growth and reproduction. This highlights the pressing need to comprehend and mitigate the effects of climate change on the distribution of invasive species.

A team of researchers at Queen's University in Canada investigated the development of adaptability in *Lythrum salicaria* by growing plants native to 25 different origins in standardized greenhouse conditions.

In the trials, gradual variations related to the original location were observed in plants, affecting flowering time and inflorescence size. Under standardized conditions, plants originating from the North started flowering earlier than those derived from southern populations. The inflorescence size was smaller in the northern individuals than in *Lythrum*'s populations native to southern regions. A similar result has also been observed in field studies of established North American populations of the purple loosestrife.

The early onset of purple loosestrife in northern populations resulted in lower seed production compared to other populations, thereby reducing the species' reproductive potential. The study showed that the decrease in seed production is not due to hereditary differences in flowering time or the size of the inflorescences.

Over the past 150 years, *Lythrum salicaria* has demonstrated remarkable adaptability, developing a series of progressively evolving characteristics (*Cline*, in biological terminology) that are well-suited to the diverse conditions of the New Continent. This adaptability, a prerequisite for survival in varying climatic conditions, is a tangible proof of the species' resilience.[59]

Lythrum salicaria does not poison its competitors

Plants forming monocultures often achieve dominance through chemical warfare, i.e., by poisoning the soil or surrounding environment with compounds that prevent the germination of seeds from other species or the development of their root systems. Such a phenomenon, known as *allelopathy*, has also been evaluated as a means of competition for *Lythrum salicaria*, which aggressively colonizes new territories.

A Japanese experimental study investigated the effects of potentially toxic chemicals released into the environment by *L. salicaria* and other riparian plants, including the emergent reed plant *Phragmites japonica* and the broadleaf cattail *Typha latifolia*.

In addition to assessing each plant alone, plantings with several species in mixtures were studied as controls. Under these experimental conditions, *Lythrum salicaria* showed no signs of allelopathy. Seeds of other plants germinated even better when grown with the purple loosestrife than with any other reference species.[60]

Distant crossbreeds increase vitality

Breeding between individuals of different origins creates new genetic combinations that are often biologically more advantageous than those of their parent forms. Genetic diversity enhances a species' ability to survive in changing environmental conditions. Usually, crossbreeding improves the biological competitiveness of a species against other species. This highlights the importance of biodiversity in maintaining ecological balance.

The effects of breeding distant relatives on the vitality and reproduction of *L. salicaria* were experimentally investigated at the University of Tübingen in Germany. The study targeted populations of diverse origins, including those from the native European habitats and North America. New gene combinations created through crossbreeding have improved growth and plant vitality, a phenomenon known as *heterosis*.

The result was anticipated, given the extensive history of *Lythrum salicaria*'s presence in its original European ranges and the New World. Having established itself in America a century and a half ago, the purple loosestrife has undergone several stages of domestication in various locations. This has led to the development of crossbreeds between different strains, thereby enabling heterosis in natural populations.

However, breeding between strains improved the vitality of the purple loosestrife only in cases where the reproducing parties were from separate origins. No improvement was observed in crossbreeds between strains spread over a vast geographical area and of the same origin.

The long history of *L. salicaria* on the New Continent has already increased the diversification of gene combinations. The crossbreeding between populations originating from different regions and at different times continues to strengthen heterosis. Therefore, the vitality of the alien plant can still improve due to the development of breeding within the species.[61]

New generations improve the genome

The genetic diversity of a plant population that has been in an established location for a long time can be diluted, meaning that the variations promoting adaptation to environmental changes will decrease. Entering new genetic material from outside the community, such as long-distance seed transport or intentional seedling transfers, benefits the species.

The characteristics of European *Lythrum salicaria* crossbreeds were studied in a series of controlled experiments conducted in populations adapted to various environmental conditions. After two successive generations, no notable deterioration in hereditary traits was observed. On the contrary, new gene combinations brought positive characteristics to the species, as expected under the principles of the natural selection theory.

The growth and reproduction characteristics of the offspring of the purple loosestrife originating from distant regions were more vital than those of any established population. The growth and seed production of the shoots were enhanced the further away or the more different conditions the interbreeding plants came from.[62]

CHAPTER 4

EXCEPTIONALLY FAST DISPERSAL

Massachusetts 1831, Louisiana 2018

The spreading pace of *Lythrum salicaria* in North America has been much faster than that of most non-native plants occupying similar habitats. For example, the narrow-leaf cattail (*Typha angustifolia*) and *T. glauca*, a hybrid between this and the broadleaf cattail (*T. latifolia*), have advanced from the east coast in a century and a half "only" into the central parts of the continent. On the other hand, the purple loosestrife made its way to the west coast of California and the southern parts of British Columbia, Canada, in just over 60 years.

According to U.S. Department of Agriculture files, *L. salicaria* was first recorded on the east coast of Massachusetts in 1831. The journey across the mainland was fast, as the newcomer was first registered in California as early as 1895. The conquest is still progressing. The most recent state where purple loosestrife was reported was Louisiana, where it was first observed in 2018.[63]

The rapid spread is primarily explained by the high seed production and the species' remarkable ability to regenerate vegetatively from pieces of rhizomes or shoots. Much-debated climate warming benefits the purple loosestrife, and an essential factor is the change in soil characteristics.

A study by Professor Susan Galatowitsch's team at the University of Minnesota revealed that the dispersal of wetland newcomers, such as *Typha glauca* and the reed canary grass (*Phalaris arundinacea*), is primarily promoted by hydrological changes resulting from habitat water management. On the other hand, the common reed (*Phragmites australis*) and narrow-leaf cattail (*Typha angustifolia*) – also alien taxa – have benefited more than representatives of native American species from changes in environmental nutrients and salinity.

The primary explanations for the exceptional success of *L. salicaria* are the absence of herbivory, i.e., the avoidance of plant-eating consumers, and the crossbreeding that resulted in new favourable gene combinations between populations from different starting points.[64]

Propagation along highways

The rapid spread of *Lythrum salicaria* is facilitated by its high seed production and human management. The expansion of suitable habitats for the species, particularly along highways, has significantly contributed to its continent-wide conquest. This highlights the importance of effective management practices in preventing the unintentional spread of invasive species and underscores the need for robust conservation efforts.

Douglas Wilcox, a researcher with the U.S. Fish and Wildlife Service, investigated the spread of *L. salicaria* across upstate New York in an east-west direction along the edges of the New York State Thruway, from Albany to Buffalo. The populations of the newcomers were mapped in ditches on both sides of the road, and on a broader scale, uniform stands were registered near the highway.

The survey revealed that the densities of established purple loosestrife communities varied in proportion to the distance from the East Coast, specifically from the initial locations of domestication for the alien species. The location of individual occurrences and the distances between them indicated that the species' progression from east to west occurs in short steps. Many moist depressions are suitable for the species along the highway, but the newcomer plant appears to have expanded into new areas "from puddle to puddle."

However, the plant's tiny seeds are believed to be capable of travelling long distances to new habitats. The extensive and lush *Lythrum* colonies in the Montezuma National Wildlife Refuge, south of Lake Ontario, facilitate local dispersal along the highway's cross-country corridor.

In the ditches of highways, it is likely easier to control a newcomer plant advancing in narrow strips toward the west than in most other suitable habitats of purple loosestrife. According to Wilcox, mowing and weeding out entire plants from the ground would be the most effective way to slow down the unwanted spread of the alien.[65]

The role of roadside ditches and the area underneath was crucial in mapping the existing stands of purple loosestrife and the spread of alien species along northern New York State and Quebec highways. The local populations of *Lythrum* along the route are small compared to the species' uniform communities occupying natural wetlands.

The local dispersal and propagation of *Lythrum salicaria* may occur relatively slowly, but this should not lead to complacency. Active management in artificial structures has been suggested as a crucial tool to prevent further invasive plant infestations. The study by researchers from the State University of New York highlights the need for intervention to control its spread, emphasizing that prompt action is necessary to prevent further spread.[66]

The ability to survive in water

The roots and rhizomes of aquatic and riparian plants are usually out of reach of atmospheric oxygen reserves. To avoid running out of oxygen, which is vital for biological functions, the plant must be able to transfer oxygen. A few plant taxa have additional or aerial roots, facilitating oxygen uptake and delivery to underground organs. Many plants growing in wet sites have anatomical structures – specialized cells in their stems and rhizomes – that facilitate the transfer of oxygen.

Lythrum salicaria has a unique feature that helps overcome the risk of oxygen debt. The purple loosestrife has cork cells, an anatomical structure commonly occurring only in woody species. The cork layer, located immediately beneath the surface of the stem and roots, consists of dead cells with numerous wide cavities that allow air to flow from the atmosphere to the underground rhizomes and roots.

A Canadian experimental study investigated how interruptions in the uniform air transport channel of the cork tissue affect the growth of purple loosestrife.

The cork layer affects the morphological structure of *L. salicaria* and the size of the plant, which certainly influences the competition between species. If the cork layer is damaged, growth in height and weight will decrease, and the proportion of biomass in the underground parts compared to the mass of the aerial shoot will also decrease.

The air transport mechanism secures the plant's growth potential even during prolonged floods when many competing species that usually thrive in the same habitats perish due to a lack of oxygen.[67]

Tree shading is a problem

The millions of dollars spent on controlling purple loosestrife have done little to slow down or limit the newcomer's ability to conquer wetlands on the New Continent. The uprooting of young shoots, mechanical cutting of emergent shoots, and attempts to poison them with pesticides have repeatedly proven disappointing. On the contrary, the natural insect enemies of *Lythrum salicaria*, especially the beetles of the genus *Galerucella*, introduced from Europe as biological control agents, have successfully reduced the local presence of invasive vegetation.

Nature also has simple means and rules to maintain a balance among diverse biota. Shading, especially from the trees that line the waterways, successfully suppresses at least six harmful non-native plants in U.S. nature. The role of environmental conditions in determining the vitality of four alien riparian plants was evaluated under the direction of Robert Warren, professor of biology at New York State University.

Surveys in the Upper Niagara River basin focused on the contribution of shade caused by trees to the growth of purple loosestrife (*Lythrum salicaria*), garlic mustard (*Alliaria petiolata*), reed canary grass (*Phalaris arundinacea*), and common reed (*Phragmites australis*). The total area of the plant populations and the seed production of each species were analyzed in relation to the extent of shading.

The researchers analyzed 348 riparian meadow sites in 12 river basins. In addition to the plant species analyses, each site's water management (hydrological) and soil and topographical (geomorphological) characteristics were also considered. The ability of an alien species to establish new habitats and expand its population is typically assessed by its seed production or vegetative reproductive capacity.

Contrary to expectations, differences in reproductive efficiency did not explain the success of newcomer plants. The seed production of all the studied plants was so high that the conquest of new territories through sexual reproduction would be possible in any part of the river basins.

Colonizing alien species and the successful, permanent conquest of new territories require favourable conditions. Riparian meadows are well-watered, and the soil in these areas is generally nutrient-rich. Therefore, the environmental factors most commonly guiding plant community structure cannot explain newcomers' success.

Based on comparisons, Warren's team showed that shading, caused by trees bordering riverside wetlands, is more critical for alien species than other measurable environmental factors. Trees are thus a positive element for the well-being of native natural biota and water quality protection. River and lakeshore restoration projects have traditionally focused on maintaining and increasing the number of trees and shrubs. Strengthening the permanent vegetation buffer zone along waterlines efficiently prevents nutrient runoff from entering water bodies.[68]

Shared characteristics and origins of unwanted aliens

The tall macrophytes *Phragmites australis* and *Lythrum salicaria*, which humans have intentionally or unintentionally transported, have become significant problems in the New World. In their areas of origin, the common reed and the purple loosestrife are widespread in the natural flora. Having settled on a new continent, both species have become expensive targets to combat and eradicate. The native ranges of *Phragmites* and *Lythrum* in Europe span over the entire continent; however, significant variations exist in the shoot morphology, population structure, and genetics of these plants. In the early 2000s, Czech researchers investigated the genetic background of these two species and whether the regions of origin may impact their success in North America.

In samples collected from the south-north gradient of Europe, the researchers found a marked correlation between the location and the plants' structural characteristics and growth potential. For both species, the original habitat explained the length of the plant's shoot, stem thickness, and underground and aboveground biomass. Shoots of *Phragmites* and *Lythrum* grew more prominent and vital in their southern habitats.

The study of the history of alien species is significant, as it has yielded a crucial finding: the morphological characteristics of the common reed and the purple loosestrife were similar in the strains native to southern Europe and in the plants that subsequently colonized North America. This data on origin provides a vital background for considering ways to control the biological invasion of alien species. Several successful projects have managed unwanted alien species with the help of the plant's natural enemies, underscoring the importance of understanding the origins of non-native organisms in ecological management.[69]

A long-term study of the invasive *Phragmites australis* in several US ponds demonstrated the significant role of the alien organism's origin. The newcomer's impact on vegetation cover and/or species richness varied between newly colonized and established reedbed communities. This variation highlights the significance of genetic variation in an organism's capacity to adapt, particularly when expanding into new territories. The study's findings on the diversity and evenness of plant communities were significant and enlightening in newly vegetated ponds and the early stages of succession.[70]

The vitality of *Lythrum salicaria* is genetically determined, and a wide variation in the genome significantly enhances the species' adaptability in nature, especially in harsh conditions. Genetic flexibility is a marked advantage that enables the establishment of new plant communities. A study by Lithuanian scientists on the genetic variation of the purple loosestrife along 15 populations in three river basins revealed the plant's impressive genetic diversity and adaptability.

Significant genetic diversity was found in *L. salicaria* within the studied populations. The larger the geographic distance between the growing sites, the greater the population variation was. However, the variation was wider within the distinct populations than between the units from different locations. In the genetic comparisons, the gene pool of the purple loosestrife was shown to consist mainly of two gene admixtures. The comparisons revealed marked differentiation in the genetic structure of *L. salicaria* between populations from distinct river basins.[71]

Salvation for pollinators in dry seasons

Lythrum salicaria, often considered an unwanted weed, is crucial in supporting pollinators, particularly in agriculture. The wildflowers in local wetlands, especially in arid areas at the end of the growing season, serve as a vital food source for pollinators when few other flowering plants are available in nature. This ecological contribution of the purple loosestrife is worth appreciating and is essential for maintaining biodiversity.

A team led by Stefano Benvenuti, a researcher at the University of Pisa in Italy, conducted a significant study on the number of invertebrate animals – primarily insects – visiting purple loosestrife and the amount of nectar or pollen they gather from the flowers. This study provides valuable insights into the plant's role in supporting pollinators.

The flowering period of the purple loosestrife is long, spanning from mid-summer to late autumn, and the large, dense spike inflorescence consists of hundreds of individual flowers. According to an Italian study, the inflorescences of *L. salicaria* in the Mediterranean region's agricultural landscapes contain an average of 640 flowers between July and September. Each flower produces a generous amount of nectar and pollen, making the purple loosestrife an excellent food source for insects that, during their visits, simultaneously pollinate the plant.

In a follow-up study spanning an entire summer, Italian researchers analyzed 476 insect pollinators that visited the purple loosestrife. The study included representatives of three orders, seven families, and 15 species. Microscopic images of insects collected from flowers found that each visitor had caught an average of 57 pollen grains from each purple loosestrife plant during the visit. Most pollinators were hymenopterans (bees and bumblebees, comprising 427 specimens), followed by numerous dipterans (flies, comprising 26 specimens) and butterflies (23 specimens).

The absolute and relative numbers of hymenopterans, a group of insects that includes bees and wasps, peaked in late summer, accounting for nearly 95 per cent of all pollinators. The honeybee (*Apis mellifera*) was the most common visitor to *L. salicaria*. Five different species of bumblebees (*Bombus* spp.) were found in the study. Visits by bees to purple loosestrife were concentrated in August, while visits by bumblebees and other insect species were uniform throughout the summer.

Due to its long flowering period and abundant nectar and pollen, the purple loosestrife is an excellent food source for insects. Italian researchers recommended planting it in agricultural areas near wetlands, especially on land used by bee-breeding farms, offering a solution to enhance biodiversity and support pollinators.[72]

Non-native organisms benefit each other

The success of *Lythrum salicaria* on the New Continent is a result of the shared origin of the plant and its essential pollinator insects. The most important insects visiting the purple loosestrife in Europe and North America are honeybees and many species of bumblebees. This mutualistic relationship between the plant and the hymenopteran insects, both of which are non-native to America, is a fascinating example of how alien organisms can benefit each other.

The purple loosestrife arrived with people in the mid-1800s, whereas the foreign hymenopterans had conquered the West much earlier. The honeybee migrated from Europe to the eastern parts of the United States in the early 1600s, and several bumblebees also arrived from the Old Continent in North America at least a hundred years before the arrival of purple loosestrife.

A collaboration of alien species – *mutualism*, a relationship where both species benefit – can help newcomers develop more robust and successful populations than the native species in the area.[73] Bees and bumblebees are the main but by no means the only pollinators of *L. salicaria*. A thorough review of American history reveals that the seed production of purple loosestrife would undoubtedly have been high even without the hymenopterans that arrived from Europe in ancient times.[74]

The usefulness of coexistence between insects and newcomers was discovered as early as the 1940s when *L. salicaria* became more common. At that time, beekeepers and honey producers began planting and spreading purple loosestrife on riverbanks and other wetlands to serve as a nectar source for honeybees.

Climate sets the limit for conquests

The rapid dispersal of *Lythrum salicaria* on the North American continent does not appear to have elicited any resistance from the local biota, and insect pests introduced from the species' native range in Europe do not seem to significantly hinder the progress of the alien species. However, the newcomer still appears to have its sensitive point: Although the purple loosestrife is found in almost all of Europe, in the New World, the spread of the species towards the North may be hindered by climatic factors.

The short growing seasons in the North significantly reduce plants' growth potential and reproductive capacity to the point where the limit for conquest may already be in sight. An experimental study investigating the possibilities and requirements for the success of purple loosestrife revealed that, on its current northern borders, the plant remains small and produces significantly fewer seeds than in its core areas. Reductions in size and seed production are adaptations by which a species reacts to external climatic requirements.

The team, led by Robert Colautti, a researcher at the University of Toronto in Canada, studied the adaptation of *L. salicaria* to the variability of the growing season. The data comprised the American range of the purple loosestrife, i.e., a 1200-kilometre south-to-north gradient from Maryland in the United States to Ontario, Canada.

The experiments revealed that purple loosestrife must adapt to a shorter growing season, with changes in shoot length and flowering schedules, as it spreads northward. The further North the plant species spread, the shorter the shoots of *L. salicaria* remain. Small-sized shoots also develop fewer flowers than taller southern individuals.

Due to the short growing season, the purple loosestrife flowers earlier the further north the plant has dispersed. Reduced size and earlier flowering lead to reduced seed production, which can be crucial in regulating and finally stopping the species' northward spread.[75]

CHAPTER 5

EFFECTS OF HERBIVORES

Escaping enemies enables enhanced growth

Living in an environment without needing to defend against plant-eating animals is the primary explanation for the purple loosestrife's exceptional success in North America. The absence of herbivorous beetles has been proven, but the plant's reactions to enemy-free conditions have been primarily based on speculations for a long time.

The hypothesis assumed that the non-native plant could give up investing resources in defence or resistance mechanisms when establishing itself in an enemy-free new environment. The energy and resources thus spared could then be used to enhance growth and reproduction. Such a new resource allocation should give the species clear advantages in competition with native plants.

The theory of optimal resource allocation was further developed into a hypothesis of the *evolution of increased competitive ability* (EICA) by a 1995 study conducted by Swiss-German researchers. To demonstrate in practice that a non-native species evolves an enhanced competitive ability in the absence of its natural enemies, the history of *Lythrum salicaria* in North America was used as an enlightening example. In the experimental study, plants were raised from seeds collected from Switzerland (where *L. salicaria* is native and exposed to herbivores) and New York State (where *L. salicaria* is non-native and has no herbivores). The vitality of the plants developed from the seeds was estimated based on growth (biomass production) and the performance of herbivores.

The results suggested that the immigrant, non-native plants in New York had lost their herbivore resistance (because the defence was unnecessary) and consequently reallocated the spared energy toward enhanced shoot growth.[76, 77]

Plant-eating animals change the flowering schedule

Spreading in the North American wild without biological enemies and nowadays classified as a severe nuisance, *Lythrum salicaria* remains in check in its native ranges, such as in Europe, where the plant has natural enemies, herbivorous animals. No herbivores are focusing on the purple loosestrife in the New World. Among the animals that feed on purple loosestrife, the most notable are the beetles of the genus *Galerucella*. These insects in the leaf beetle family have been transported from the Old Continent to the US to prevent the non-native plant from establishing and dispersing.

The effects of herbivorous animals typically extend to the entire plant, affecting growth in height, biomass production, flowering, and seed production. In the above-ground shoot, the damaged leaves and stems weaken the plant's vitality and decrease production. In addition, the shoot eaters indirectly cause damage by changing the plant's appearance. By eating leaves, herbivores reduce the plant's attractiveness to other animals. Pollinators may not be interested in a damaged plant, even if it has numerous flowers and offers a generous supply of nectar and pollen.

Direct and indirect effects are exerted by herbivorous animals operating beneath the ground level. Herbivores that eat the root system and rhizomes reduce the plant's total biomass and simultaneously worsen the physiological condition of the individual plant. The most severe consequence of herbivory is reducing or inhibiting the plant's flowering, thereby decreasing seed production, even if pollinator visits are not affected in this manner.

An experimental study of competitive relationships between *L. salicaria* and the agriculturally important grass Timothy (*Phleum pratense*) specifically demonstrated the disadvantages of underground herbivores coexisting with purple loosestrife. In a study by researchers from Christian-Albrechts-University in Germany, *Hylobius transversovittatus* beetles were added to the soil in the root zone of plants. These herbivorous insects, which belong to the weevils, damage the hair roots vital to plant physiology and responsible for water and nutrient uptake.

The tiny invertebrate animals slowed down and reduced the growth of *L. salicaria*'s shoots in the first growing season and reduced biomass production in the second year of impact. However, the most severe consequence of weevil feeding was the change in the schedule and effectiveness of flowering. The purple loosestrife bloomed later than usual due to weevils' underground herbivory, and the generally abundant-flowered spike developed fewer flowers than normal.

Damage to the plant's roots and rhizomes caused by weevils weakened *L. salicaria* more than the competition with Timothy. In Timothy, the length increment decreased markedly as the weevils ate the plants' root mass. Simultaneously, the entire hay population began forming short, lawn-like growth. However, these changes did not prevent the normal development of the purple loosestrife in the plant community.[78]

Herbivory can increase flowering

Herbivorous animals can alter the characteristics of the plants they consume, even in ways that benefit the plants themselves.

In some cases, damages change the plants' growth schedule and the interrelationships between the plant organs. In non-native species, the role of herbivores is more complicated than in native flora. Damages to leaves, stems, and inflorescences caused by the larvae and adults of beetles feeding on *Lythrum salicaria* can change the size proportions of the inflorescences, as well as the flowering schedule.

Researchers at the University of Ottawa in Canada investigated the effect of purple loosestrife's natural (caused by leaf beetles of the genus *Galerucella*) and artificial (mechanical cutting of leaves and stems) damage on the inflorescences and pollination of the target plant's flowers. The experiments were conducted in natural populations of purple loosestrife under uniform environmental conditions. During the trial, shoot growth, inflorescence size, and flowering time were measured. The effect of herbivores on pollinator visits was studied at the end of the growing season, when the larval stage of the leaf beetles had completed their feeding and pupated.

Experiments revealed an increase in the number of inflorescences and flowers due to the activity of herbivores. Regardless of the activity of plant feeders and the mass of tissues lost, the target plant always produced enough pollen to guarantee pollination and fertilization in undamaged and damaged plants.

The flowering schedule also changed. As a result of shoot damage caused by herbivores, the flowering of the purple loosestrife began later than in the undamaged plants.[79]

Bumblebees are attracted to damaged plants

The effects of herbivory on the behaviour and outcome of pollinating insects visiting *Lythrum salicaria* were investigated in Canada by researchers at the University of Ottawa. Perhaps surprisingly, plants damaged by leaf-eating beetles proved to be favourites to pollinators.

Bumblebees visited the inflorescences that grew on shoots previously damaged by beetles even more. The pollination efficiency was enhanced by insects visiting flowers in inflorescences exposed to herbivores more frequently and for extended periods than in the undamaged control group plants.

Changes in inflorescence structure or pollinator behaviour caused by herbivory did not affect the reproductive output of *L. salicaria*. At the end of the growing season, no differences were found in the number of seeds developing between treatments.

However, the flowering schedule changed by the activity of herbivores can affect the resistance of the purple loosestrife. Throughout the long-term coevolution, the life cycles of herbivores have synchronized with those of the plants. Thus, if herbivores alter the flowering schedule, the activity period of most plant-eating larvae may not coincide with the optimal flowering time in purple loosestrife.

Changes in a plant's shoot caused by herbivores can slow growth and production and significantly alter flowering and seed production. If eating the shoot alters the flower's structure, quantity, or flowering schedule, the changes will undoubtedly affect the behaviour of potential pollinators. Such changes were experimentally studied in North American populations of the purple loosestrife.

When studying the effects of herbivores, purple loosestrife plants were exposed to beetles. Black-margined loosestrife beetles (*Galerucella calmariensis*) and their relatives, the *G. pusilla* beetles, have been used in the United States for the biological control of the non-native invasive species. In the experiments, the species and activity of pollinators visiting the flowers of purple loosestrife were studied at a stage of the growing season when the herbivory of beetle larvae was ending.

As a main result, it was found that more flowers developed on shoots eaten by beetles than on shoots that avoided pest attacks. However, the flowering schedule changed due to herbivores. On the shoots eaten by the larvae, the flowers opened significantly later than in other *Lythrum salicaria* plants of the same population.

The behaviour of pollinators visiting the damaged plants differed markedly from that of the same insects in other individuals within the studied populations of *L. salicaria*. Bumblebees are essential pollinators of purple loosestrife, and their activity was abnormal in flowers damaged by herbivores. In damaged plants, bumblebees visited more flowers than average and flew between different parts of the inflorescence more actively than the average for *L. salicaria* spikes.[80]

Beetles invite other insects to forage on purple loosestrife

In nature, it is common for herbivorous animals to focus on only one or a few plant species in their foraging. *Lythrum salicaria* has specialized exploiters, the most common and best known being the beetles of the genus *Galerucella*, members of the leaf beetle family (Chrysomelidae).

Like many other herbivores, leaf beetles gather in suitable forage plants in large groups – both in the larval stage and as adults. Beetles actively promote group formation by secreting chemical decoys, i.e., *pheromones*, which are species-specific or closely related to their own. Beetles are further attracted to plants due to the chemical compounds released from the tissues of *L. salicaria* that herbivores damage.

Peter Hambäck, a researcher at the Department of Botany at Stockholm University, studied the diverse chemical interactions occurring when leaf beetles injure (eat) *Lythrum* tissues. The analyses of pheromones showed that leaf-eating beetles released hormones into the environment. The compounds secreted by the purple loosestrife's damaged cells attract a broad group of herbivore species, injuring the plant and its close relatives.

The chemical interactions in the lives of pheromone-attracted insects are complex and intricate. The hormone analyses revealed that *Galerucella pusilla* is more attracted to the purple loosestrife flowers by chemical messages sent by the Black-margined loosestrife beetle (*G. calmariensis*) than by compounds released through herbivory by another individual of *G. pusilla*. Beetles tend to gather in dense groups on the target plant. This could be expected to intensify competition for food resources. However, the leaf mass of *L. salicaria* is sufficient for all parties, regardless of the number of insects involved.

Cooperative mechanisms are profitable for the foragers. Laboratory and field experiments showed that the larvae of beetles acting in groups grow faster than plant-eating insects working alone.

Cooperation between closely related beetle species damages plants more than the herbivory of a single species. Thus, the group behaviour of herbivores should be considered when planning management procedures in the biological control of non-native, harmful species.[81]

Herbivores get along with each other

The two most often used herbivores to control *Lythrum salicaria* only marginally compete for food or space. Researchers at Virginia Polytechnic Institute and State University in the United States compared, under controlled experimental conditions, the numbers and growth rates of larvae of *Galerucella calmariensis* and *G. pusilla* leaf beetles, as well as the amounts of leaf mass consumed by the insects. The effects of temperature on herbivory were also assessed in tests conducted at five temperatures ranging from 12.5°C to 27.5°C (54.5°F to 81.5°F). At temperatures above 30 degrees, larval development ceased, and the beetles failed to develop into adults.

The amount of plant mass the beetles eat is central to the success of biological control of the purple loosestrife. Black-margined loosestrife beetle (*G. calmariensis*) larvae were, on average, 25% more effective than their relatives. Among the adult beetles, *G. pusilla* consumed more than its relative at temperatures between 15°C (59°F), and 27.5°C (81.5°F), whereas in cooler conditions, the consumption caused by *G. calmariensis* was higher.

Competition between leaf beetles is crucial in determining the intensity of herbivory only when the edible plant mass is scarce. For biological control, beetles are introduced to destroy established populations. Therefore, the domestication of either species of *Galerucella* yields the desired results.[82]

Plant-eaters are selective

In the New World, there are several close relatives to the purple loosestrife, even plants belonging to the same genus, *Lythrum*. These plants could be assumed to be subject to beetles introduced as biological control tools. An experiment led by University of Minnesota researcher Elizabeth Stamm Katovich investigated the reactions and effects of two of the most famous *Galerucella* beetles on the purple loosestrife, as well as on two other plants belonging to the same Myrtaceae family (winged loosestrife, *Lythrum alatum* and the water willow, or swamp loosestrife, *Decodon verticillatus*).

Breeding experiments investigated the effects of leaf beetles on the growth and seed production of the three plants. The foraging of leaf beetle larvae and adults was studied at different stages of plant development. Plants grown under uniform conditions and not exposed to herbivores were compared.

European leaf beetles, which function as biological control agents for *Lythrum salicaria* in North America, reduced seed and biomass production more in the target plant than in other plant species. The most apparent effect of herbivory was the reduction in the height increment of the purple loosestrife. In two species related to *L. salicaria*, the influence of beetles was minimal or completely non-existent.

For biological control to be successful, the development cycles of plants and their herbivores must coincide during the growing season. This timing was investigated by comparing the flowering schedules of the purple loosestrife and two other plant species with the natural life cycle of leaf beetles.

The active developmental phase of the beetles coincided with the most sensitive period for the growth and flowering of the purple loosestrife. In contrast, the local native species *L. alatum* begins flowering a month later than *L. salicaria*. Due to the one-month time difference, the beetles are no longer active leaf-eaters in the flowering phase of the plant. Thus, thanks to the "temporal escape", *L. alatum* avoids the damage caused by herbivores, even though the leaves are suitable for beetles.[83]

Aliens are strong

Established evolutionary theories suggest that liberation from enemies enables plants to gradually lose the unnecessary ability to defend themselves against herbivores.

Researchers at the University of Tübingen in Germany investigated the phenomenon's realization by exposing two populations of *Lythrum salicaria* to herbivores originating from the plant's traditional range. In experimental conditions, purple loosestrifes derived from Europe were compared to plants from the species' new territory in North America.

Plants used in experiments came from (a) seeds collected from plants growing in wild populations (both European and American origins) and (b) from seeds derived from the first-phase experimental plantations.

In accordance with the theory, experiments confirmed the weakening of the plant's resistance. The damage to *Lythrum salicaria* was more significant in American individuals than in European individuals, both in absolute terms and in terms of the plants' total leaf biomass. The damage caused by herbivores slowed down the growth of the American-borne purple loosestrife. However, their shoots grew taller than their European counterparts by the end of the growing season.

The production of shoots showed that the tolerance of the purple loosestrife against herbivorous insects is more critical than the genetically determined defences. Reactions to herbivores were more pronounced in second-generation plants than in individuals grown directly from seeds collected from the wild.

The results show that the competitive advantage and ability to spread, resulting from liberation from natural enemies, are hereditarily determined characteristics of *L. salicaria*. The acquired characteristic is preserved even under new, enemy-free conditions.[84]

Resistance changes as required

The biological control of *Lythrum salicaria*, based on the domestication of beetles, presents difficulties due to the considerable variation in life cycles and reactions to environmental conditions between the plant and the insects that feed on it. Since the range of purple loosestrife is extensive, climate, soil, and other environmental conditions also exhibit significant variations.

Sweden offers excellent opportunities to study the adaptation of *Lythrum salicaria* and *Galerucella* beetles to varying environmental conditions. The plant is found naturally throughout the long Baltic Sea coastline, and leaf beetles thrive on this 1200-kilometre south-north axis.

Ecologists Lena Lehndal and Jan Ågren from the University of Uppsala studied the resilience of the purple loosestrife to changes in the abiotic environment and natural stressors, as well as the plant's ability to resist damage caused by leaf beetles. The study's 12 follow-up sites spanned a geographical area of 10 latitudinal degrees (56°N – 66°N) along the east coast of Sweden, extending from southern Skåne to the northernmost shores of the Gulf of Bothnia. Due to the broad range, climatic conditions vary significantly at different research points.

The length of the growing season varies from 210 to 220 days at the southernmost research points to 140 to 150 days for sites in the North. The average annual temperature ranges from 7°C to 8°C (44.6°F to 46.4°F) in the south to 1°C to 2°C (33.8°F to 35.6°F) in the north.

The field studies were complemented by experiments conducted in standardized greenhouse conditions, which investigated the mechanisms of purple loosestrife under various beetle exposure alternatives.

The study aimed to determine whether *L. salicaria* reacts differently to damage caused by beetles under different climatic conditions. Two types of resistance were investigated: the plant's capacity to withstand varying environmental conditions and stress (tolerance) and its ability to defend against biological pests, such as the feeding on tissues by *Galerucella* leaf beetles (resistance).

It was previously established that the growth and flowering schedules of *L. salicaria* vary significantly across different coastal regions. On the other hand, it was also known that the number of beetles decreased from south to north.

The purple loosestrife's tolerance to stressors in the abiotic environment was better in the northern locations than on the south coast. The differences are at least partly due to the timing throughout the growing season: The populations in the North, because of the earlier onset of growth, are older than the plants in the southern populations at the critical stage when leaf beetle invasion and plant-eating begin in the summer.

The study found that *Lythrum salicaria's* resistance to herbivores decreases as habitats shift from south to north, with shorter growing seasons. However, the plant's tolerance to abiotic stress in the environment improves under these conditions.[85]

In another study, the Uppsala researchers demonstrated the central role of herbivory in determining the demographic cycle, as well as the course and intensity of flowering and seed production, in purple loosestrife. The effects of plant-feeding invertebrates were experimentally studied over a two-year period by artificially removing all herbivores from plants and comparing the vitality of these plants with that of neighbouring plants.

Along the vast latitudinal gradient, herbivory was strongest in the southern plant communities, and the role of feeders decreased linearly towards the North. At the southern site, herbivores consumed 11 per cent of the *Lythrum's* leaf area, whereas leaf removal was only 3 per cent in the northernmost population.

L. salicaria individuals grew 1.6 times taller in the southern population, where the herbivores were experimentally removed. Further north, the height increase was 1.2-fold.

The removal of plant feeders also increased fertility. In the southern population, herbivore-free individuals had four times more flowers, and seed production was 1.6 times higher in the herbivore-free population. In Central Sweden, flowering was twofold, and seed output was 1.2 times higher due to the removal of herbivores. In the northernmost population, the artificial removal of herbivores had no effect on the fitness or fertility of the plants.

The study concludes that native herbivores play a significant role in shaping the demographic structure of *Lythrum salicaria* populations, thereby influencing the geographic patterns of seed production. The strength of herbivory is geographically determined, with a more substantial impact on plant fitness in the south and a decreasing effect towards the north.[86]

Herbivores are the main inhibitors of seed production

The experimental study by the Swedish universities of Uppsala, Stockholm, and Umeå investigated the relationships between *Lythrum salicaria* and plant-eating beetles in nine coastal populations in the northern Swedish archipelago. The aim was to examine the crucial role of herbivores in the growth and seed production of purple loosestrife, as well as the environmental conditions that affect plant populations.

To assess the state of the plants, the studied *L. salicaria* populations were selected so that environmental stress caused by sea ice, waves, and water level fluctuations resulted in varying degrees of stress on the plants. In controlled breeding experiments, differences in resistance to pests were measured among individuals from each population of different origins.

Most often, small herbivorous insects – nearly invisible to the human eye – slow down shoot growth and limit flowering and seed production. The biomass production of purple loosestrife increased 500 times, and the number of flowers and seeds increased 400 times under conditions in which herbivores specializing in *L. salicaria* were removed. In European natural habitats, these invertebrates are always found where the purple loosestrife grows.

Plant damage caused by beetles was the lowest among the populations, which were most affected by environmental stress from ice and waves. The contribution of herbivory to the flowering and seed production of the purple loosestrife was studied by removing all beetles from plants. Removing the beetles demonstrated that herbivores were the primary factor in reducing seed production.

Based on the experiment, the differences in flowering intensity and seed production among plant populations were almost exclusively due to herbivorous animals. The abiotic stress factors had only a marginal effect on the success of *L. salicaria* in these northern seashore habitats.

However, experimental studies have shown that, despite significant variation in outcome, the genetically determined resistance of purple loosestrife to herbivores varies substantially between individuals within each population.[87]

Biological control weapons escaped into gardens

The beetles introduced to America to feed on *Lythrum salicaria* were believed to be safe in their new habitats. On the European side, *Galerucella* beetles have not caused problems with cultivated or ornamental plants; therefore, these herbivores have been introduced to many destinations in North America. However, the plantation of living organisms always comes with potential risks, and such a risk materialized in Oregon.

As a joint project of the State Department of Agriculture, the U.S. Department of Agriculture, and the City of Portland, black-margined loosestrife beetles (*Galerucella calmariensis*) were released into the nature conservation area known as Oaks Bottom Refuge to control the purple loosestrife.

This biological control had been ongoing for years, and the beetles successfully fulfilled their mission. However, in August 2015, residents were surprised as thousands of small beetles suddenly invaded their yards, gardens, and homes. Experts alerted to the scene confirmed that the intruders were beetles that had been deliberately released into the wild by the authorities.

Nothing like this had happened before. Experts of the ministry explained that the extraordinary reproduction of beetles and population expansion led to the invasion of beetles into settlements and gardens, where herbivorous larvae destroyed roses and other garden plants. Scientists estimated that the unusually hot and dry summer resulted in an explosive increase in the beetle population.[88]

CHAPTER 6

SUCCESS IN SEVERE ENVIRONMENTS

Advanced heavy metal tolerance

The ability of plants to withstand harmful exposures varies significantly, and many species possess unique properties that enhance their tolerance. These capabilities, coupled with physiological adaptation mechanisms, enable growth and well-being even in a toxic environment, highlighting the impressive adaptability of nature.

Species with such characteristics can conquer and dominate in problematic environments and often form strong monocultures. *Lythrum salicaria* appears to be one species that can tolerate extremely high concentrations of heavy metals, even if those elements are toxic to most other plants.

The ability of the purple loosestrife to tolerate lead (Pb) entering the wild from human activities was studied at Cleveland State University in the United States. In the experimental conditions, the lead content of the plants' rooting substrate was (a) Low: 500 milligrams per litre (mg/l), (b) Medium: 1000 mg/l, and (c) High: 2000 mg/l. A level rated high indicates it is toxic for most organisms. Control plants were grown in lead-free soil.

The effects of lead were harmful to plants under all experimental conditions. Even in the least lead-exposed plants, the height growth, weight development, and number of leaves of purple loosestrife remained lower than those of control plants. However, all test plants survived, even at the highest lead exposures, although their growth was significantly reduced at a lead concentration of 2000 mg/l.

Based on the experiments, the tolerance to heavy metals may be one factor that enables the purple loosestrife's ability to conquer areas contaminated by human activities.[89]

Tolerant to road salt

The harmful properties of soil or water can inhibit the germination of plant seeds. The seeds of aquatic and riparian plants often end up in sludge and sediments, which, in addition to essential nutrients, are enriched with harmful chemicals.

The biota of coastal zones is often stressed by chemical loads carried by snowmelt waters every spring. The stress is at its worst near heavily fertilized agricultural lands, industrial operations, and residential areas. During snowmelt, water collected from streets carries many pollutants, including those from car exhaust emissions and road surfaces, into the watersheds. A special danger to living biota is the salting of roads. Especially in inland coastal zones, the spring pulse of salty meltwaters can burden biota.

The impacts of snowmelt on five species of riparian plants were experimentally investigated in Ontario, Canada. A mixture of seeds was sown and watered with roadside snowmelt, diluted snowmelt or tap water. Within one month, the growth of all five species slowed, biomass production declined, and the wetland plants became impoverished. The adverse effects were directly proportional to the amount and concentration of snowmelt.

In conditions modelling exposures to high salt concentrations in waters draining from the roadways, only

Lythrum salicaria and the broadleaf cattail (*Typha latifolia*) survived. Due to the limited tolerance to high sodium chloride exposure, the diversity of local plant communities is expected to decrease. The loss of native flora in such challenging conditions enables the purple loosestrife to establish and gain dominance.

However, despite its endurance, the purple loosestrife also suffered from the effects of salty waters. When the irrigation water was collected entirely from the street, only 9% of the *L. salicaria* seeds germinated. Germination, of course, is never 100%. In ordinary irrigation water, with no chemicals in the meltwater, the seed germination rate of purple loosestrife was 30%, and in the broadleaf cattail (*Typha latifolia),* only 13% of the seeds developed. The seeds of any other wetland plant did not germinate under the influence of street meltwaters.[90]

Works even in wastewater purification

A study by Nanjing University in China revealed the effectiveness of *Lythrum salicaria* as a nutrient collector and binder. The experiments compared the ability of the purple loosestrife and the sweet lag or swag (*Acorus calamus*) to accumulate nutrients from artificial wastewater. The solution conducted through the plant's substrates contained 80 milligrams of nitrogen per litre, significantly more than conventional urban wastewater.

In an eight-month experiment, wetland plants collected and stored 35–63% of the nitrogen, 47–76% of the phosphorus, and 22–40% of the organic carbon in the liquid's nutrients. Both plant species grew and remained in good condition at all concentrations of the artificial wastewater administered.

The nutrients passing through the wetland plant stands were quickly bound and accumulated in plant tissues throughout the experiments. The enrichment of nutrients proved the usefulness of wetland plant communities as natural tools for water purification. In nature, the communities of wetland plants bordering the shores are priceless in water protection.

Among the plant species studied, *Acorus calamus* proved to be a more effective binder of nitrogen nutrients, whereas the tissues of *L. salicaria* accumulated phosphorus more efficiently.[91]

Another experimental study conducted in China obtained even more promising results regarding the ability of *L. salicaria* to accumulate nutrients and thus reduce environmental eutrophication. Using impressive plants, the study demonstrated that technical and aesthetic goals can be achieved simultaneously and at reasonable costs. In comparing three wetland macrophytes – *Lythrum salicaria*, *Sagittaria trifolia*, and *Typha orientalis* – the purple loosestrife proved to be the most effective in removing nitrogen and phosphorus nutrients from simulated contaminated waters.

During the 35-day experiment, the purple loosestrife removed 87–97% of the total nitrogen and 46–67% of the total phosphorus from the nutrient-rich water. The supremacy of the purple loosestrife was based on the high biomass and abundance of microorganisms in the plant's rhizosphere.

The Chinese scientists' study concluded that *Lythrum salicaria* and other emergent wetland plants are useful in urban and rural landscape design, reducing eutrophication and assisting in phytoremediation.[92]

Using the ability of vegetation to bind nutrients is a standard practice in wastewater treatment. Tall emergent riparian plants, such as cattails (*Typha* spp.) and common reed (*Phragmites australis*), are widely used worldwide. Still, other plants are also suitable for sequestering nutrients and harmful chemicals from dirty water. A study by Yulin University in China investigated the resilience of five plant species to wastewater derived from a steelworks coke oven.

All plants absorbed ammonium and oxygen-consuming materials from diluted coke oven wastewater drained through the vegetation. Biological treatment is most successful when plants are used in tertiary treatment plants, i.e., to treat further wastewater that has already been treated by other methods before being discharged into a water body.

Of the five plant species compared, *Lythrum salicaria* was one of the most effective accumulators of harmful substances. The species also had excellent tolerance to the chemical load of wastewater. However, when raw sewage accounted for more than 30% of the total flow, the plants' physiological functions deteriorated due to the high concentrations of harmful substances.

Exposure to wastewater was a stress to plants used in purification. The most apparent effects of stress were the deterioration of photosynthesis and the inhibition of essential enzyme activity. Despite minor drawbacks, the best candidates, such as *L. salicaria*, proved to be valuable purifiers of coke oven wastewater that severely burdens water bodies.[93]

Chapter 7

Control by various means

Unpredictable results with poisons

In agriculture, various chemicals of different compositions and modes of action are used to fight unwanted weeds. Attempts have also been made to use pesticides (poisons) to attack *Lythrum salicaria*, which spreads wildly in North America. However, the results do not seem promising, as the control's effectiveness and permanence are inconsistent.

Tests conducted over 10 years at the University of Nebraska in the United States studied the efficacy of 14 pesticides in controlling the growth of purple loosestrife. The pesticides used (each in two different potencies) were glyphosate, 2,4-D dimethylamine, trichloropyr, imazapyr, metsulfuron, fosamine, a combination of trichloropyr and 2,4-D amine, and a combination of metsulfuron and 2,4-D amine.

The best results were achieved using high doses of glyphosate, imazapyr, and metsulfuron. These three pesticides reduced purple loosestrife by over 90%, and the desired impact lasted 360 days. Fosamine and 2,4-D amine were the least effective against *L. salicaria*.

The effectiveness of the chemical treatment depended on the pesticides used and the age of the purple loosestrife plants. The older and more established the weed population was, the more intense the chemical treatments needed to destroy the plant.[94]

Poisoning may even advance the target

Chemical control can eradicate *Lythrum salicaria*, but the treatment may give the alien plant even better chances to thrive than if it were left intact. The most widely used pesticide, glyphosate, does not destroy the hard-shelled seeds of the purple loosestrife.

A barren and pesticide-free substrate is a favourable base for developing new seedlings. Hence, *L. salicaria*, with its well-known enormous seed production, often has seed banks in the soil that are ready for germination and colonization of suitable habitats. Unsuccessful glyphosate treatment may not eliminate purple loosestrife but instead create favourable conditions for the non-native coloniser.

If the purple loosestrife is controlled with glyphosate, the treatment should be supplemented by applying 2,4-D, which prevents seed germination.

Weeding out the alien is difficult

Mechanical uprooting of entire plants is the simplest method for controlling the spread of colonizing non-native plants. Mechanical management of *Lythrum salicaria* may be possible if the plant is identified soon after it has been established. However, mowing seldom stops the plant's spreading due to the various ways the purple loosestrife can propagate.

Changing agricultural practices have dramatically altered vast grasslands, especially in Western societies. The cessation of cattle grazing has led to a decline in natural herb communities, and consequently, the colonization of strong competitors. To stop the degradation of biodiversity, mowing is a natural way to protect the natural or endemic phytocenoses. Mowing often works in the short term. However, controlling strong invaders – whether or not the local vegetation was previously kept in check by cattle – requires repeated cuttings, often over several consecutive years.

The success of repeated mowing in controlling vital grassland plants is unstable. While one invader is successfully controlled, the other can survive and thrive despite yearly cuttings. A comprehensive Polish study has shown that purple loosestrife appears to be one of those persistent survivors.

Researchers Aleksander Kołos and Piotr Banaszuk at Białystok University of Technology experimentally studied the effects of mowing on the growth and persistence of three wetland macrophytes in two river valleys within the Białowieża Forest. The performance of the yellow loosestrife (*Lysimachia vulgaris*), the purple loosestrife (*Lythrum salicaria*) and the meadowsweet (*Filipendula ulmaria*) was followed for 15 years under various management practices.

The management of the purple loosestrife stands was hard to predict. *L. salicaria* coped regardless of mowing intensity, mainly when grown on dry and seldom flooded hummocks. The year-to-year variations were significant in the plant's location and the sizes of separate populations. To be successful, the mowing procedures must be done before flowering. If the plant can produce seeds, the seed banks in the habitat's topsoil keep the alien viable practically endlessly.[95]

Many sides to the alien's nuisance

Lythrum salicaria, a striking, purple-flowered newcomer, is one of North America's most harmful non-native organisms. Its deliberate planting and spreading are prohibited, and significant resources are annually allocated to control its spread. The plant's impressive appearance during flowering has contributed to its positive image, but the original biodiversity of nature does not celebrate its exceptionally rapid spread. To effectively manage the invader, holistic approaches are needed. The evaluations should include an expansive approach to specify the broader ecological context and directly analyse the plant.

In just over a century and a half, the purple loosestrife conquered all the continental states of the United States except Florida. Elizabeth Farnsworth, a researcher at the New England Wild Flower Society, and Donna Ellis, a researcher at the University of Connecticut, asked, twenty years ago, in the scientific journal *Wetlands*, whether *Lythrum salicaria* should be classified exclusively as an invasive species. The extensive range of the newcomer and the locally achieved dominance do not differ significantly from those of the native species. The question is primarily which variables are used in comparisons.

To clarify the dilemma, the researchers studied the position of *L. salicaria* in five wetland habitats in Connecticut. At each site, 30 plots of one square metre were analyzed. The density of each plant species (number of individuals per unit area) and the shoot biomass at the end of the growing season were examined using standard methods in plant ecology. All plant species growing on the plots were identified and listed.

The relationship between the native plants in the plots and the purple loosestrife was analyzed using three variables: (a) Total plant production, i.e., *L. salicaria* biomass versus biomass of other species; (b) Species diversity, i.e., the number of individuals of the purple loosestrife versus the total number of other species, (c) Vegetation coverage, i.e., the total amount of the individuals of the purple loosestrife versus the coverage of other plants.

Although the tall and spectacular *Lythrum* often stands out in wetlands and coastal meadows as uniform, single-species monocultures, many native plants thrive in non-native newcomer communities. In the study, neither the number of individuals of the purple loosestrife nor the coverage of its shoots was correlated with the reduction of other plant species.

The frequency of occurrence of other plant species (i.e., the percentage of all squares in which the species is found) did not significantly depend on the density of purple loosestrife. In other words, the average occurrence of other plants was the same regardless of whether the density of Lythrum salicaria was low or high. The conclusion regarding the diversity or presence of plant taxa in Lythrum-occupied wetlands was that "no other plant species was unequivocally associated with an alien species, and no single species can be said to have been unequivocally lost due to the presence of purple loosestrife."

However, as the total number of plant taxa is examined, the purple loosestrife does change the flora of wetlands. When comparing the standing crops of plants, the biomass of other species is negatively correlated with the biomass of L. salicaria. In other words, as the populations of the alien species became more extensive and denser, the numbers and biomass of other plant species decreased.

On plots where L. salicaria was the dominant species, the biomass of the alien was significantly higher than that of other plant taxa. In contrast, in stands where other native plants dominated, regardless of the presence of L. salicaria, the total biomass of the native plants was the greatest. In the long term, all the subjugated species will inevitably suffer due to competition and reduced reproductive potential.

Evaluations of the status and effects of Lythrum salicaria require analyses of several ecological variables, not only the presence or absence of a species in a habitat. Studying the biomass and production of plant species provides essential information on the mechanisms by which non-native species conquer and achieve dominance in wetland communities.[96]

Other newcomers to be considered in alien control

The extensive monocultures of *Lythrum salicaria* can dominate the landscape, and the populations, despite their beauty, are not desirable. The success of non-native plants has deteriorated the status of native biota in several regions worldwide. At worst, the intense competitor can irrevocably displace the local biota. The purple loosestrife is a strong competitor but is only one of many successful newcomers occupying the wetlands.

Growing in moist meadows and bordering waterlines in North America, the reed canary grass (*Phalaris arundinacea*) – a common non-native species – is almost as effective in conquering new fields as the purple loosestrife. Researchers from Oregon State University and the U.S. Department of Agriculture compared the success and competitiveness of the two newcomers in 24 wetlands in the northeastern Pacific states. The fates and successes of the strangers have several uniform features.

The study analyzed four characteristics of species diversity in wetland vegetation. In addition to the number of species per area, three mathematical indices of plant communities' diversity were studied.

Field studies revealed that the effects of the two newcomers on vegetation diversity were consistent across all four characteristics measured. A vital feature explaining the newcomers' success and determining the plant community's well-being was the density of the conqueror's shoots, which refers to the number of individuals per unit area.

The success of *Lythrum salicaria* and *Phalaris arundinacea* is mainly due to their unique characteristics. Local site conditions, such as moisture, soil quality, or topography, have influenced the effectiveness with which non-native species have displaced native flora.

The uniform competitive set-up of the two alien species highlights how difficult it is to fight strangers. In the United States, millions of dollars have been invested in eradicating and controlling *L. salicaria*. However, more resources are needed. Suppose the goal is to safeguard the native flora of wetland meadows and manage purple loosestrife. In that case, attention should also be paid to other invasive species, such as reed canary grass. The researchers emphasize that blocking one unwanted species could lead to another newcomer gaining dominance in the local flora.[97]

Control measures to be combined

In the wetlands of Canada's St. Lawrence Delta, *Lythrum salicaria* is the most common invasive plant species. However, a stranger dominating the landscape, especially during the flowering period, is not the most significant newcomer in total numbers and effects. In a study of non-native organisms led by Queen's University researcher Keiko Lui, the densities of at least two other alien species and their adverse competitive effects were found to be greater than those of purple loosestrife.

The common reed (*Phragmites australis*) and reed canary grass (*Phalaris arundinacea*) displace more native wetland species in the St. Lawrence Delta than the purple loosestrife (*Lythrum salicaria*). However, only the purple loosestrife is subject to hate and active control.[98]

In rare cases, the fight against the invasive purple loosestrife has been successful as expected or desired. Eradicating a non-native species by only one means, even though theoretically effective, is usually insufficient in natural conditions. A series of experiments conducted by the University of Manitoba in Canada demonstrated that combining biological control with a chemical pesticide treatment can be an effective solution.

In three-year field trials on two-hectare *L. salicaria* populations, the alien plant was dozed with glyphosate alone in one set. In another series, the stands of purple loosestrife were exposed to different amounts of *Galerucella calmariensis* leaf beetles. In the third series of experiments, leaf beetles were transferred to the glyphosate-treated *Lythrum* stand. In each series of experiments, control efficacy was evaluated by measuring the plant's height, the size of the inflorescences, the number of flowers, the number of fruits produced by the flowers, and the total number of individuals in the community.

The integrated treatment combining chemical and biological control significantly reduced the number of *L. salicaria* shoots on the plots. When acting alone, *Galerucella* leaf beetles did not reduce the number of shoots, but the growth of the plants decreased by almost 70%. Chemical control, based solely on glyphosate treatment, was insufficient to eradicate the alien species. In contrast, the number of shoots of the purple loosestrife increased during the experiment.[99]

The use of herbicides may or may not affect the occurrence of the alien purple loosestrife, but the choice to poison wildlife must always be based on detailed information. Unsuccessful weed controls have all too often led to devastating damage to the environment.

CHAPTER 8

ALIEN PLANTS IMPACT THE ANIMAL KINGDOM

Some birds suffer, but many benefit

Non-native plants have effectively colonized new territories, displacing native plants and forming monocultures in various habitats worldwide. However, there are only a few studies on the effects of aggressively reproducing newcomers on the structure of the rest of the biosphere. Birds can quickly move to other habitats, but the emergence of a new dominant plant community can lead to changes in feeding and nesting opportunities, for better or worse.

Studies in the US have found that establishing *Lythrum salicaria* has significantly altered the composition of local bird populations. Several bird species, especially those in coastal wetlands, were observed to avoid nesting and feeding in areas dominated by purple loosestrife.[100]

Brian Tavernia, a researcher at the National Audubon Society and Tufts University researcher J. Michael Reed compared the composition of birdlife in wetlands with varying environmental conditions and plant species. A special target of observation was how the presence of the purple loosestrife and the density or size of the plant populations affected the birdlife. The generally quoted assumption that a non-native plant invader's effects are categorically adverse appeared too straightforward. Still, many bird species suffered due to the change in flora caused by purple loosestrife.

Several bird species also benefited and became more abundant after *L. salicaria* established itself in the wetlands. No unambiguous model of the reactions of the alien plant and bird species emerged in the comparisons. Local conditions vary within any swamp ecosystem, so undeniable cause-and-effect relationships between the purple loosestrife and birdlife could not be established.

The wide variation from positive to negative makes it challenging to evaluate *Lythrum salicaria's* general effects in North American nature. The researchers conclude that some birds benefit from the newcomer, and this should be considered when assessing the significance of the ecological changes caused by purple loosestrife.[101]

Invading plants change the diet of animals

The quantity and quality of food significantly influence the physiological condition and performance of animals, and local supply plays a crucial role in nutrition. The diet of herbivores can change radically if a new plant species with a different chemical composition than the established local flora colonizes a habitat.

The effects of introducing the non-native purple loosestrife on the well-being of amphibians were investigated in the United States. In two mesocosm microhabitat experiments, tadpoles of two amphibian species were fed with locally derived hardwood detritus and whole *Lythrum salicaria* plant tissues. The food selection and use by the frog larvae were followed, and the role of the diet was evaluated. The target animals were the wood frog (*Rana sylvatica*) and the southern leopard frog (*R. sphenocephala*).

Dietary experiments on the larvae were conducted outdoors, and the amphibian tadpoles were simultaneously offered a mixture of local plant material and decomposed and decaying cells of purple loosestrife. The food quality consumed by the tadpoles of frogs and the proportions of different plant species in their diets were investigated by analyzing the isotopes of nitrogen (^{15}N) and carbon (^{13}C) separated from plant materials.

The results showed that the larvae of both frog species actively select *Lythrum salicaria* as their food source. The result can be surprising, as the tissues of the purple loosestrife are known to contain compounds harmful to animals, such as tannins, in significantly higher concentrations than most local plant materials. This may be because the tissues of the purple loosestrife also contain several beneficial nutrients more than plant detritus on average.[102]

In another experiment, the larvae of the American toad (*Bufo americanus*) were affected by compounds dissolved in the litter of purple loosestrife. Still, under similar conditions, the compounds released from another emergent wetland plant, the broadleaf cattail (*Typha latifolia*), did not affect the development of the larvae. The negative results appeared to be species-specific, as the larvae of *Hyla versicolor* tree frogs grown under the same conditions were not affected by the compounds of the purple loosestrife or cattails. Scientists conclude that various tannins and phenolic compounds are harmful compounds released from decaying plant tissues.

The sensitivity of the American toad larvae to tannins released from *Lythrum* tissues was attributed to the respiratory mechanism of the larvae. In these frogs, oxygen supply is based solely on gills, and tannins were estimated to have damaged these organs. On the other hand, *Hyla* tree frogs and young larvae breathe through their lungs throughout their lifespan.

The lungs are a more resilient organ than the gills in tolerating exposure to secondary compounds in plants.[103]

In previous experiments elsewhere, the detritus of purple loosestrife has been shown to slow the development of young American toads. Still, no adverse effect was observed with cattail's (*Typha* sp.) detritus.

Analyses of the intestinal contents of the larvae showed that the species composition of small-sized algae differs in the waters dominated by *Lythrum salicaria* and *Typha latifolia*. The change in algae-containing diets was the reason for the growth retardation of toad larvae and young individuals, as well as increased tadpole mortality.

Indirectly, through changes in the species composition of algae and due to the quality of the litter from vascular plants (such as purple loosestrife and broadleaf cattail), the structure and function of the food webs within the wetland community change, which in turn impacts the functioning of the entire ecosystem.[104]

Sheep keep the alien in check

The eutrophication of nature, which has progressed rapidly in recent decades, is reflected in dramatic changes to water bodies and riparian meadows. In many places, the dominant species of wetlands bordering both seashore and inland waters, especially the common reed (*Phragmites australis*), have been managed to prevent the degradation of biodiversity. A biological method of controlling non-desired invaders is grazing, which is done by cattle or sheep. This natural method is also effective in controlling *Lythrum salicaria*.

In a 2011 study, researchers at New York State University found that sheep in a riparian meadow preferred to eat *Lythrum salicaria* at various stages of the plant's growth cycle. Thus, sheep can control non-native plants that, if not managed by human efforts, often grow rapidly and colonize habitats.

In an experimental study conducted in wet riparian meadows, sheep were kept for one growing season, from May to August 2008, in small enclosures, with the animals' sites changed every two to three days. Control sections of the meadow were inaccessible to grazers.

Grazing sheep consumed the purple loosestrife and other meadow vegetation so effectively that most plants failed to develop to the flowering stage. The coverage of the central target plant, *L. salicaria*, was reduced by an average of 40.7% due to grazing. No changes in plant coverage were noted in the ungrazed meadow sections.[105]

Changes in the species composition of a meadow ecosystem indicate the strong biological competitiveness of *Lythrum salicaria*. In areas grazed by sheep, the diversity of meadow vegetation increased by 20% due to the activity of plant feeders compared to the species composition of the reference areas.

Wild mammals and sheep use *L. salicaria* as their food crop, and several animal species even prefer it. The non-native plant is included in the diet of at least white-tailed deer, muskrats, hares, and rabbits. In addition, some large bird species – for example, the red-winged blackbird (*Agelaius phoeniceus*), one of North America's most abundant bird species – are known to use purple loosestrife as a food source.

CHAPTER 9

MEDICINES AND COSMETICS

Traditional remedy in folk medicine

The external beauty of *Lythrum salicaria* is undoubtedly acknowledged, but the plant also offers remarkable values in its inner beauty. Various tissues of the purple loosestrife contain biologically active compounds, many of which humans have used for centuries.[106]

Historical sources vividly illustrate the extensive use of *L. salicaria*, particularly in Europe, as a medicinal or herbal plant effective against various ailments. Many sources describe the herbal use of purple loosestrife in treating gastrointestinal disorders, particularly diarrhoea and dysentery, as well as stopping bleeding. Externally, *L. salicaria* has been used to cure skin diseases, a tradition passed down through generations.

Before the breakthrough of synthetic preparations, the purple colour obtained from the flowers of *L. salicaria* was used to decorate pastries and sweets. The rhizomes of the species are rich in tannins, and these compounds have been used by fishers in the Black Sea and Caspian Sea to strengthen the threads of their nets and prevent unwanted creatures from sticking to the gear.

Besides its medical use, the active compounds of purple loosestrife are suitable for skin care and enhance the healing of skin lesions.

Even in modern times, the potential of *L. salicaria* continues to be explored, with new applications being discovered. A recent experimental study by researchers from French cosmetic companies and the University of Orléans unveiled the effectiveness of ellagitannins extracted from *Lythrum salicaria* in strengthening skin structures. These ellagitannins, abundant in berries and fruits, are flavonoids known for their medicinal effects, particularly in cancer treatment. This discovery opens up a world of possibilities for the new commercial applications of purple loosestrife in the pharmaceutical and cosmetic industries, highlighting its economic significance.

The study, led by researcher Glorianne Jouravel, investigated the effects of compounds extracted from the shoots of purple loosestrife on the properties of the most common skin cells, keratocytes. Additionally, plant extracts are beneficial in promoting cell renewal in the epidermis, the outermost layer of the skin.

The tannins of purple loosestrife were also found to improve the properties of grafts used to treat injured skin. In addition to the cellular and tissue levels, positive effects were demonstrated in caring for healthy, undamaged skin. The plant also has potential as a raw material for modern cosmetics.[107]

Lythri herba: Versatile, traditional healer

While the historical use of the purple loosestrife for medicinal purposes is well-documented, its current use is limited. However, the plant's effectiveness has not diminished, and there is a pressing need for an updated assessment of its potential uses. *Lythrum salicaria*, included in the catalogue of medicinal plants as *Lythri herba*, is a subject that not only warrants but also excites further research and exploration.

A team of researchers from the Medical University of Warsaw in Poland compiled published data on the medical use of the purple loosestrife over several centuries. Based on numerous indications and descriptions, scientists characterize purple loosestrife as an exceptionally effective remedy in traditional folk medicine. Additionally, the publication highlights the therapeutic potential of the compounds found in various parts and cells of the plant.

Although *Lythri herba* is currently of little use, the chemical properties of the purple loosestrife have also been studied extensively using modern analytical methods. Indeed, information on the biologically active compounds of the plant can be considered accurate. The main compounds suitable for herbal use are found in a broad group of polyphenols. Ellagitannins and flavonoids, commonly found in berries and fruits, are utilized in medicinal and herbal preparations. Pharmacologically active compounds are also present in several polysaccharides (sugars) isolated from *Lythrum salicaria*.

Based on chemical analyses, the compounds obtained from *L. salicaria* have antimicrobial, antioxidant, anti-inflammatory, and antidiabetic effects.[108]

In further experiments at the Warsaw University of Medicine, the positive effects of the purple loosestrife were primarily directed at the white blood cell neutrophils. In these tests, the primary impact of the plant extracts was to eliminate or attenuate inflammatory conditions.

The positive impacts of ellagitannins were demonstrated in the functions of metabolically important carbohydrates and vital enzymes.[109]

The trust in the healing powers of the purple loosestrife remains alive and well even today. Several suppliers sell powders, creams and liquids prepared from extracts derived from *Lythrum salicaria* online.

Drug residues may even enhance plant growth

The use of chemicals, including pharmacologically active ingredients in medicines, is harmful to biota when discharged into the environment. Drug residues accumulate in food webs, but the actual effects of drugs on the organisms treated in nature's ecosystems are poorly understood. There are studies on the effects of drug residues on animals, but little is known about the reactions of plants.

Researchers at the Tor Vergata University in Rome, Italy, studied the impacts of the antibiotic drug sulfadimethoxine on *Lythrum salicaria*. For four weeks under experimental conditions, the effects of the drug on *L. salicaria*'s growth and the accumulation of drug residues in the plant's different organs were monitored. The plants were exposed to drug doses with a 10,000 times concentration range.

The effects of the drug on purple loosestrife depended on the dose of chemicals, and the results varied across different plant organs. In very low concentrations, the antibiotic increased the growth of the stems and leaves of *L. salicaria*. In high doses, the drug residues were toxic to the plant. The roots and seedlings of purple loosestrife are affected by the antibiotic properties in all doses, including the lowest concentrations.[110]

CHAPTER 10

NEWCOMER BECOMES PART OF NATURE

Native species can adapt to an alien

The vitality of *Lythrum salicaria* on the New Continent is an undeniable competitive stress factor for native biota. This competitive stress, which includes resource competition and habitat alteration, can lead to the displacement of other species, especially during the newcomer's colonizing period. However, the long-term ecological impacts of the purple loosestrife may be relatively minor, as it eventually becomes an established part of wetlands.

In a topographic study conducted by researchers from the University of British Columbia and the University of Guelph in Canada, the effects of purple loosestrife proved significant in the years following the species' colonization. The study found that the newcomer's dominant position has weakened over time. Decades after its establishment, the purple loosestrife has become, at least in some regions, a regular part of wetlands, allowing other typical riparian species to return to their habitats.

Along the Bar River in Ontario, vegetation in 41 wetland areas was studied, with a focus on the proportion of purple loosestrife within the overall vegetation. Contrary to the general assumption that invasive species negatively impact plant diversity, the diversity of the plant kingdom remained uniform regardless of whether the alien species occurred or not.

The number of plant species decreased when *L. salicaria* inhabited the habitat in the monitored areas. The numbers of many typical Canadian wetland species, including several species of bulrushes (*Scirpus* spp.) and the yellow pond-lily or variegated pond-lily (*Nuphar variegata),* increased as they grew together with *Lythrum salicaria.* The study concluded that the spread of the European newcomer had not reduced the number of species of wetland plants in Ontario.[111]

Similar results were obtained in the study in central New York State. Researchers from Cornell University compared plant community dynamics and ecosystem properties in an abandoned agricultural field and a pristine wetland, where *Lythrum salicaria* had been invaded at both sites. No changes were observed in the communities' total biomass, growth, or biomass of the native vegetation, or soil properties, due to the establishment of purple loosestrife.

In conclusion, the *L. salicaria* invasion did not seriously hurt the biota in this relatively pristine wetland.[112]

Even natural flora has detrimental conquerors

Alien species that reproduce and spread aggressively are a real threat to natural biota. Still, the importance of non-native organisms can be exaggerated, at least in some places. In the Canadian province of Ontario, researchers at the University of Ottawa investigated the origins of tall, dominating plant species found in the vegetation of 58 wetlands.

Four native species typical of the area were observed: the broadleaf cattail (*Typha latifolia*), the narrow-leaf cattail (*Typha angustifolia*), the willow species *Salix petiolaris,* and the yellow pond-lily (*Nuphar variegata*), as well as four non-native species, the purple loosestrife (*Lythrum salicaria*), the Eurasian frogbit (*Hydrocharis morsus-ranae*), the reed canary grass

(*Phalaris arundinaceae*) and the deciduous tree, alder buckthorn (*Rhamnus frangula*).

In the study areas, the non-native alien plants did not achieve dominance or adversely displace other plants more often than native Canadian plants. The researchers' conclusion and recommendation were to observe communities critically, even when any species, whether a representative of native nature or an alien, emerges as dominant.[113]

The impacts of colonizing non-native macrophytes are typically characterized as detrimental, but this is not always the case. A comprehensive meta-analysis by researchers at the University of Plymouth of 202 case studies on the effects of alien plants on recipient habitats' plants, invertebrates, and fish emphasized that the species-specific characteristics of each invader must be studied in detail; no generalized conclusions can be drawn.

The global analysis of the impacts of alien plants does not explicitly mention *Lythrum salicaria*, suggesting that the threat posed by this species is primarily local. Among the tall emergent plants colonizing the same habitats as the purple loosestrife, the most serious competitors are various species and hybrids of the cattail genus (*Typha* spp.) and the "most widely distributed macrophyte in the world," the common reed (*Phragmites australis*).

In most cases, the impacts of *Lythrum salicaria* and other alien macrophytes on natural biodiversity are detrimental. However, changes in species composition do not necessarily lead to a decline in the community's productive capacity or total biomass.[114]

The complexity of areal and temporal variability in the effects of non-native intruders on the local biota is described and verified in hundreds of studies worldwide and is briefly summarized in this book. Therefore, it is not surprising that the position and appreciation of *Lythrum salicaria* cover the entire spectrum in nature's diversity.

Even in the well-studied and recognized explanation for the remarkable vitality of the purple loosestrife's success in new territories, i.e. the escape from natural enemies, there are numerous side shows. The many factors affecting the alien intruders' effects – sometimes positive, sometimes disastrous – are comprehensively described in the review of the *Enemy Release Hypothesis* in aquatic and riparian plants, published in 2025 by South Africa's Rhodes University researchers.[115]

There are no definitive answers to the question of *Lythrum salicaria*'s effects in nature. The plant is an alien in most of its current destinations, and in new territories, the intruder has displaced local plants and animals. But at the same time, several organisms get along well with the newcomer, and some even benefit from it. As the expert in wetland ecology, Professor Arnold van der Valk at Iowa State University emphasizes that wetland assemblages dominated by purple loosestrife often differ from communities dominated by native species, but does it really matter?[116]

References

1   Darwin C. 1859. *The Origin of Species.*

2   IPBES (Intergovernmental Platform on Biodiversity and Ecosystem Services). Roy HE, Pauchard A, Stoett P & Renard TT (Editors). 2023. *The Thematic Assessment Report on Invasive Alien Species and their Control*; https://www.ipbes.net/ias

3   Prosser RS & Brain RA. 2024. Where have all the flowers gone? A systematic evaluation of the factors driving the decline of native terrestrial plants in North America. *Environmental Science and Pollution Research* 31: 48460-48483; https://doi.org/10.1007/s11356-024-34349-9

4   van der Valk AG. 2012. *The Biology of Freshwater Wetlands.* Second Edition. Oxford University Press, Oxford.

5   Hajek A. 2004. *Natural Enemies. An Introduction to Biological Control.* Cambridge University Press, Cambridge.

6   Macêdo RL, Haubrock PJ, Klippel G, Fernandez RD, Leroy B, Angulo E, Carneiro L, Musseau CL, Rocha O, Cuthbert RN. 2024. The economic costs of invasive aquatic plants: A global perspective on ecology and management gaps. *Science of The Total Environment* 908: 168217; https://doi.org/10.1016/j.scitotenv.2023.168217

7   Allaby M. 2019. *Oxford Dictionary of Plant Sciences.* Fourth Edition. Oxford University Press, Oxford.

8   Vilizzi L. *et al.* (39 authors). 2025. To be, or not to be, a non-native species in non-English languages: gauging terminological consensus amongst invasive biologists. *Management of Biological Invasions* 16(1): 15–31; https://doi.org/10.3391/mbi.2025.16.1.02

9   Silvertown J. 2005. *Demons in Eden. The Paradox of Plant Diversity.* The University of Chicago Press, Chicago & London.

10 Fleming JP & Dibble ED. 2015. Ecological mechanisms of invasion success in aquatic macrophytes. *Hydrobiologia* 746: 23–37; https://doi.org/10.1007/s10750-014-2026-y

11 Anderson MG. 1995. Interactions between Lythrum salicaria and native organisms: A critical review. *Environmental Management* 19(2): 225–231; https://doi.org/10.1007/BF02471992

12 Thomas V & Faircloth N. 2014. *Shakespeare's Plants and Gardens. A Dictionary.* Bloomsbury, London.

13 Cao L, Larson J & Sturtevant R. 2024. *Lythrum salicaria* L.: U.S. Geological Survey, Nonindigenous Aquatic Species Database, Gainesville, FL; https://nas.er.usgs.gov/queries/FactSheet.aspx?SpeciesID=239

14 Torres CD & Puntieri JG. 2015. *Lythrum salicaria* (Lythraceae), nueva cita para la flora de Argentina (*Lythrum salicaria* (Lythraceae), new record for the Argentinean flora). *Darwiniana, nueva serie* 3.2 (2015): 208–213; https://doi.org/10.14522/darwiniana.2015.31.668

15 Geerts S & Adedoja O. 2021. Pollination and reproduction enhance the invasive potential of an early invader: the case of *Lythrum salicaria* (purple loosestrife) in South Africa. *Biological Invasions* 23: 2961–2971, 2021; https://doi.org/10.1007/s10530-021-02549-w

16 New Zealand Plant Conservation Network. 2025. Lythrum salicaria; Lythrum salicaria • New Zealand Plant Conservation Network

17 McAlpine KG & Howell CJ. 2024. List of environmental weeds in New Zealand 2024; https://www.doc.govt.nz/globalassets/documents/science-and-technical/sfc340.pdf

18 Stuckey RL. 1980. Distributional history of Lythrum salicaria (Purple Loosestrife) in North America. Bartonia: *Journal of the Philadelphia Botanical Club* 47: 3–20.

19   Wilson M, Schwarzlaender M, Blossey B & Randall CB. 2004. *Biology and Biological Control of Purple Loosestrife.* USDA Forest Service / UNL Faculty Publications. Paper 112. http://digitalcommons.unl.edu/usdafsfacpub/112

20   Schmidt RJ, King MR, Aronson MFJ & Struwe L. 2023. Hidden cargo: The impact of historical shipping trade on the recent-past and contemporary non-native flora of the northeastern United States. *American Journal of Botany* 110(9): e16224; https://doi.org/10.1002/ajb2.16224

21   NAISMA (North American Invasive Species Management Association), 7.1.2025. Purple Loosestrife: An Invasive Species; https://naisma.org/2025/01/07/purple-loosestrife-an-invasive-species/

22   Global Invasive Species Database; Downloaded March 27.2025; https://www.iucngisd.org/gisd/100_worst.php

23   Lavoie C. 2010. Should we care about purple loosestrife? The history of an invasive plant in North America. *Biological Invasions* 12(7): 1967–1999; https://doi.org/10.1007/s10530-009-9600-7

24   Pimentel D, Lach L, Zuniga R & Morrison D. 2000. Environmental and economic costs of nonindigenous species in the United States. *BioScience* 50: 53–65; https://doi.org/10.1641/0006-3568(2000)050[0053:EAECON]2.3.CO;2

25   Blossey B, Skinner LC & Taylor J. 2001. Impact and management of purple loosestrife (Lythrum salicaria) in North America. *Biodiversity and Conservation* 10(10): 1787–1807; https://doi.org/10.1023/A:1012065703604

26   Mal TK, Lovett-Doust J, Lovett-Doust L & Mulligan GA. 1992. The biology of Canadian weeds. 100. Lythrum salicaria. *Canadian Journal of Plant Sciences* 72(4): 1305–1330; https://doi.org/10.4141/cjps92-164

27   Whigham DF, McCormick J, Good RE & Simpson RL. 1978. Biomass and primary production in freshwater tidal wetlands of the Middle Atlantic coast. *In:* Good RE, Whigham DF,

Simpson RL (Editors), *Freshwater Wetlands. Ecological Processes and Management Potential*, pp. 3–18. Academic Press, New York.

28   Darwin C. 1864. On the sexual relations of the three forms of *Lythrum salicaria. Journal of the Proceedings of the Linnean Society of London. Botany* 8(31) 169–196; http://onlinelibrary.wiley.com/doi/10.1111/j.1095-8312.1864.tb01085.x/full

29   Thompson K. 2018. *Darwin's Most Wonderful Plants. Darwin's Botany Today.* Profile Books. London.

30   Costa J, Castro S, Loureiro J & Barrett SCH. 2017. Experimental insights on Darwin's cross-promotion hypothesis in tristylous purple loosestrife (*Lythrum salicaria*). *American Journal of Botany* 104(4): 616–626; https://doi.org/10.3732/ajb.1600408

31   Darwin C. 1877. *The Different Forms of Flowers on Plants of the Same Species.* With a new Foreword by Herbert G. Baker, 1986. The University of Chicago Press.

32   Balogh CM & Barrett SCH. 2018. Genetic and environmental influences on partial self-incompatibility in *Lythrum salicaria* (Lythraceae). *International Journal of Plant Sciences 179*(6): 423–435; https://doi.org/10.1086/698211

33   Brown BJ, Mitchell RJ & Graham SA. 2002. Competition for pollination between an invasive species (Purple loosestrife) and a native congener. *Ecology* 83(8): 2328–2336; https://doi.org/10.1890/0012-9658(2002)083[2328:CFPBAI]2.0.CO;2

34   Denoth M & Myers JH. 2007. Competition between *Lythrum salicaria* and a rare species: combining evidence from experiments and long-term monitoring. *Plant Ecology* 191(2): 153–161, 2007; https://link.springer.com/article/10.1007/s11258-006-9232-2

35   CABI Digital Library (N. Pasiecznik). Updated 1.11.2024. *Lythrum salicaria* (Purple loosestrife). In: *Invasive Species Compendium.* Wallingford, UK: CAB International; https://www.cabi.org/isc/datasheet/31890#tosummaryOfInvasiveness

36   Soons MB, Van der Vlugt C, Van Lith B, Heil GW & Klaassen
      M. 2008. Small seed size increases the potential for dispersal of
      wetland plants by ducks. *Journal of Ecology* 96(4): 619–627;
      https://doi.org/10.1111/j.1365-2745.2008.01372.x

37   Welling CH & Becker RL. 1993. Reduction of purple
      loosestrife establishment in Minnesota wetlands. *Wildlife Society
      Bulletin 21*(1), 56–64; www.jstor.org/stable/3783362

38   Yakimowski SB, Hager HA & Eckert CG. 2005. Limits and
      effects of invasion by the nonindigenous wetland plant *Lythrum
      salicaria* (purple loosestrife): a seed bank analysis. *Biological
      Invasions* 7(4): 687–698; https://doi.org/10.1007/
      s10530-004-5858-y

39   Colautti RI, Eckert CG & Barrett SCH. 2010. Evolutionary
      constraints on adaptive evolution during range expansion in an
      invasive plant. *Proceedings of the Royal Society B. Biological
      Sciences* 277(1689): 1799–1806; https://doi.org/10.1098/
      rspb.2009.2231

40   Klips RA & Peñalosa J. 2003. The timing of seed fall, innate
      dormancy, and ambient temperature in *Lythrum salicaria.
      Aquatic Botany 75(1): 1–7;* https://doi.org/10.1016/
      S0304-3770(02)00148-1

41   Maki K & Galatowitsch S. 2004. Movement of invasive aquatic
      plants into Minnesota (USA) through horticultural trade.
      Biological Conservation 118(3): 389–396; https://doi.org/
      10.1016/j.biocon.2003.09.015

42   Stewart A. 2010. *Wicked Plants. The A-Z of Plants That Kill,
      Maim, Intoxicate and Otherwise Offend.* 236 pages. Timber
      Press, London.

43   Lym RG. (North Dakota State University). 2004. Identification
      and Control of Purple Loosestrife (*Lythrum salicaria* L.);
      https://library.ndsu.edu/server/api/core/bitstreams/71432085-
      0be9-43f4-b966-03c1b002aea6/content

44   Twolan-Strutt L & Keddy PA. 1996. Above- and below-ground
      competition intensity in two contrasting wetland plant

communities. *Ecology* 77(1): 259–270; https://doi.org/
10.2307/2265675

45   Keddy P, Fraser LH & Wisheu IC. 1998. A comparative
      approach to examine the competitive response of 48 wetland
      plant species. *Journal of Vegetation Science* 9(6): 777–786;
      https://doi.org/10.2307/3237043

46   Hovick SM, Bunker DE, Peterson CJ & Carson WP. 2011.
      Purple loosestrife suppresses plant species colonization far more
      than broad-leaved cattail: experimental evidence with plant
      community implications. *Journal of Ecology* 99(1): 225–234;
      https://doi.org/10.1111/j.1365-2745.2010.01754.x

47   Keddy P, Gaudet C & Fraser LH. 2000. Effects of low and high
      nutrients on the competitive hierarchy of 26 shoreline plants.
      Journal of Ecology 88(3): 413–423; https://doi.
      org/10.1046/j.1365-2745.2000.00456.x

48   Hager HA & Vinebrooke RD. 2004. Positive relationships
      between invasive purple loosestrife (*Lythrum salicaria*) and plant
      species diversity and abundance in Minnesota wetlands.
      Canadian Journal of Botany 82(6): 763–773; https://doi.
      org/10.1139/b04-052

49   Anderson MG. 1995. Interactions between *Lythrum salicaria*
      and native organisms: A critical review. *Environmental
      Management* 19: 225–231: https://doi.org/10.1007/
      BF02471992

50   Morrison JA. 2002. Wetland vegetation before and after
      experimental purple loosestrife removal. *Wetlands* 22: 159–169;
      https://doi.org/10.1672/0277-5212(2002)022[0159:WVBAAE
      ]2.0.CO;2

51   Mazurczyk T & Brooks RP. 2022. Native biodiversity increases
      with rising plant invasions in temperate, freshwater wetlands.
      Wetlands Ecology and Management 30(1): 139–160; https://doi.
      org/10.1007/s11273-021-09842-4

52   Edwards KR, Adams MS & Kvet J. 1998. Differences between
      European native and American invasive populations of Lythrum

salicaria. *Applied Vegetation Science* 1(2): 267–280; https://doi. org/10.2307/1478957

53   Chun YJ, Kim C-G & Moloney KA. 2010. Comparison of life history traits between invasive and native populations of purple loosestrife (*Lythrum salicaria*) using nonlinear mixed effects model. *Aquatic Botany* 93(4): 221–226; https://doi. org/10.1016/j.aquabot.2010.09.001

54   Halkka O & Halkka L. 1974. Polymorphic balance in small island populations of *Lythrum salicaria*. *Annales Botanici Fennici* 11(4): 267–270; http://www.jstor.org/stable/23725077

55   Heuch I. 1980. Loss of incompatibility types in finite populations of the heterostylous plant, *Lythrum salicaria*. *Hereditas* 92(1): 53–57; https://doi. org/10.1111/j.1601-5223.1980.tb01678.x

56   Dech JP & Nosko P. 2004. Rapid growth and early flowering in an invasive plant, purple loosestrife (Lythrum salicaria L.) during an El Niño spring. *International Journal of Biometeorology* 49(1): 26–31; DOI https://doi.org/10.1007/ s00484-004-0210-x

57   Colautti RI & Barrett SC. 2013. Rapid adaptation to climate facilitates range expansion of an invasive plant. *Science* 342(6156): 364–366; https://www.science.org/doi/10.1126/ science.1242121

58   Colautti RI, Ågren J & Anderson JT. 2017. Phenological shifts of native and invasive species under climate change: insights from the *Boechera-Lythrum* model. *Philosophical Transactions of the Royal Society of London; B. Biological Sciences* 372(1712); https://doi.org/10.1098/rstb.2016.0032

59   Montague JL, Barrett SCH & Eckert CG. 2008. Re-establishment of clinal variation in flowering time among introduced populations of purple loosestrife (*Lythrum salicaria*, Lythraceae). *Journal of Evolutionary Biology* 21(1): 234–245; https://doi.org/10.1111/j.1420-9101.2007.01456.x

60   Iriyama Y, Uchida T & Takayama M. 2004. Allelopathic activities among four hydrophytes. *Journal of Japanese Society of Revegetation Technology* 30(1): 169–174; https://doi.org/10.7211/jjsrt.30.169

61   Shi J, Joshi J, Tielbörger K, Verhoeven KJF & Macel M. 2018a. Costs and benefits of admixture between foreign genotypes and local populations in the field. *Ecology and Evolution* 8(7): 3675–3684; https://doi.org/10.1002/ece3.3946

62   Shi J, Macel M, Tielbörger K & Verhoeven KJF. 2018b. Effects of admixture in native and invasive populations of *Lythrum salicaria*. *Biological Invasions* 20(9): 2381–2393; https://doi.org/10.1007/s10530-018-1707-2

63   Cao L, Larson J & Sturtevant R. 2024. *Lythrum salicaria* L.: U.S. Geological Survey, Nonindigenous Aquatic Species Database, Gainesville, FL; https://nas.er.usgs.gov/queries/FactSheet.aspx?SpeciesID=239

64   Galatowitsch SM, Anderson NO & Ascher PD. 1999. Invasiveness in wetland plants in temperate North America. *Wetlands* 19(4): 733–755; https://doi.org/10.1007/BF03161781

65   Wilcox DA. 1989. Migration and control of purple loosestrife (*Lythrum salicaria* L.) along highway corridors. *Environmental Management* 13(3): 365–370; https://doi.org/10.1007/BF01874916

66   Rogers J, Humagain K & Pearson A. 2022. Mapping the purple menace: spatiotemporal distribution of purple loosestrife (Lythrum salicaria) along roadsides in northern New York State. *Scientific Reports* 12: 5270; https://doi.org/10.1038/s41598-022-09194-w

67   Stevens KJ, Peterson RL & Reader RJ. 2002. The aerenchymatous phellem of *Lythrum salicaria* (L.): a pathway for gas transport and its role in flood tolerance. *Annals of Botany* 89(5): 621–625; https://doi.org/10.1093/aob/mcf088

68   Warren RJ, Potts DL & Frothingham KM. 2015. Stream structural limitations on invasive communities in urban riparian areas. Invasive *Plant Science and Management* 8(3): 353–362; https://doi.org/10.1614/IPSM-D-14-00081.1

69   Bastlová D, Bastl M, Čížková H & Kvet J. 2006. Plasticity of *Lythrum salicaria* and *Phragmites australis* growth characteristics across a European geographical gradient. *Hydrobiologia* 570(1): 237–242; https://link.springer.com/article/10.1007/s10750-006-0186-0

70   Amatangelo KL, Stevens L, Wilcox DA, Jackson ST & Sax DF. 2018. Provenance of invaders has scale-dependent impacts in a changing wetland ecosystem. *NeoBiota* 40: 51–72; https://doi.org/10.3897/neobiota.40.28914

71   Jocienė L, Krokaitė E, Shakeneva D, Rekašius T, Stanys V, Šikšnianienė J B, Žvingila D, Paulauskas A, & Kupčinskien E. 2022. Relationship between genetic and environmental characteristics of Lithuanian populations of purple loosestrife (Lythrum salicaria). *Journal of Environmental Engineering and Landscape Management* 30(1), 81–93. https://doi.org/10.3846/jeelm.2022.16303

72   Benvenuti S, Benelli G, Desneux N & Canale A. 2016. Long lasting summer flowerings of *Lythrum salicaria* as honeybee-friendly flower spots in the Mediterranean basin agricultural wetlands. *Aquatic Botany* 131: 1–6; http://www.sciencedirect.com/science/article/pii/S0304377016300080

73   Simberloff D & Von Holle B. 1999. Positive interactions of nonindigenous species: Invasional meltdown? *Biological Invasions* 1(1): 21–32; https://doi.org/10.1023/A:1010086329619

74   Mal TK, Lovett-Doust J, Lovett-Doust L & Mulligan GA. 1992. The biology of Canadian weeds. 100. Lythrum salicaria. *Canadian Journal of Plant Sciences* 72: 1305–1330; https://cdnsciencepub.com/doi/pdf/10.4141/cjps92-164

75 Colautti RI, Eckert CG & Barrett SCH. 2010. Evolutionary constraints on adaptive evolution during range expansion in an invasive plant. *Proceedings of the Royal Society B. Biological Sciences* 277(1689): 1799–1806; https://doi.org/10.1098/rspb.2009.2231

76 Blossey B & Nötzold R. 1995. Evolution of increased competitive ability in invasive nonindigenous plants: a hypothesis. *Journal of Ecology* 83(5): 887–889; https://doi.org/10.2307/2261425

77 Lockwood JL, Hoopes MF & Marchetti MP. 2010. *Invasion Ecology*. Blackwell Publishing.

78 Nötzold R, Blossey B & Newton E. 1997. The influence of below ground herbivory and plant competition on growth and biomass allocation of purple loosestrife. *Oecologia* 113(1): 82–93; https://doi.org/10.1007/s004420050356

79 Thomsen CJM & Sargent RD. 2017. Evidence that a herbivore tolerance response affects selection on floral traits and inflorescence architecture in purple loosestrife (*Lythrum salicaria*). *Annals of Botany* 119(8): 1295–1303; https://doi.org/10.1093/aob/mcx026

80 Russell-Mercier JL & Sargent RD. 2015. Indirect effects of herbivory on plant–pollinator interactions in invasive *Lythrum salicaria*. *American Journal of Botany* 102(5): 661–668; https://doi.org/10.3732/ajb.1500043

81 Hambäck PA. 2010. Density-dependent processes in leaf beetles feeding on purple loosestrife: aggregative behaviour affecting individual growth rates. *Bulletin of Entomological Research* 100(5): 605–611; https://doi.org/10.1017/S000748530999068X

82 McAvoy TJ & Kok LT. 2007. Fecundity and feeding of *Galerucella calmariensis* and *G. pusilla* on *Lythrum salicaria*. *BioControl* 52(3): 351–363; https://link.springer.com/article/10.1007/s10526-006-9042-4

83   Stamm Katovich EJ, Becker RL, Ragsdale DW & Skinner LC. 2008. Growth and phenology of three Lythraceae species in relation to feeding by *Galerucella calmariensis* and *Galerucella pusilla*: Predicting ecological host range from laboratory host range testing. *Invasive Plant Science and Management* 1(2): 207–215; https://doi.org/10.1614/IPSM-07-039.1

84   Joshi S & Tielbörger K. 2012. Response to enemies in the invasive plant *Lythrum salicaria* is genetically determined. *Annals of Botany* 110(7): 1403–1410; https://doi.org/10.1093/aob/mcs076

85   Lehndal L & Ågren J. 2015a. Latitudinal variation in resistance and tolerance to herbivory in the perennial herb *Lythrum salicaria* is related to the intensity of herbivory and plant phenology. *Journal of Evolutionary Biology* 28(3): 576–589; https://doi.org/10.1111/jeb.12589

86   Lehndal L & Ågren J. 2015b. Herbivory differentially affects plant fitness in three populations of the perennial herb *Lythrum salicaria* along a latitudinal gradient. *PLoS ONE* 10(9): e0135939; https://doi.org/10.1371/journal.pone.0135939

87   Lehndal L, Hambäck PA, Ericson L & Ågren J. 2016. Herbivory strongly influences among-population variation in reproductive output of *Lythrum salicaria* in its native range. *Oecologia* 180(4): 1159–1171; https://link.springer.com/article/10.1007%2Fs00442-015-3520-2

88   International Pest Control. 2015. USA: Beetles meant to curb weeds invade neighbourhood. International Pest Control, 13 October 2015; http://international-pest-control.com/news-in-brief-sepoct15/

89   Uveges JL, Corbett AL & Mal TK. 2002. Effects of lead contamination on the growth of *Lythrum salicaria* (purple loosestrife). *Environmental Pollution* 120(2): 319–323; https://doi.org/10.1016/s0269-7491(02)00144-6

90   Isabelle PS, Fooks LJ, Keddy PA & Wilson SD. 1987. Effects of roadside snowmelt on wetland vegetation: an experimental study. *Journal of Environmental Management* 25: 57–60.

91   Zhao Y, Liu B, Zhang W, Kong W, Hu C & An S. 2009. Comparison of the treatment performances of high-strength wastewater in vertical subsurface flow constructed wetlands planted with *Acorus calamus* and *Lythrum salicaria*. *Journal of Health Science* 55(5): 757–766; https://doi.org/10.1248/jhs.55.757

92   Sun L, Liu J, Zhao H, Wang Z, Liu X, Chang Y & Yao D. 2022. Phytoremediation performance of three traditional ornamental hydrophytes and the structure of their rhizosphere microorganism populations. *Environmental Science and Pollution Research* 29: 50727–50741; https://doi.org/10.1007/s11356-022-19543-x

93   Xiang Y, Xiang Y, Wang L & Jiao Y. 2018. Effects of coking wastewater on the growth of five wetland plant species. *Bulletin of Environmental Contamination and Toxicology* 100(2): 265–270; https://doi.org/10.1007/s00128-017-2215-2

94   Knezevic SZ, Osipitan OA, Oliveira MC & Scott JE. 2018. *Lythrum salicaria* (Purple loosestrife) control with herbicides: Multiyear applications. *Invasive Plant Science and Management* 11(3): 143–154; https://doi.org/10.1017/inp.2018.17

95   Kołos A, Banaszuk P. 2021. How to remove expansive perennial species from sedge-dominated wetlands: results of a long-term experiment in lowland river valleys. *Rendiconti Lincei. Scienze Fisiche e Naturali* 32: 881–897; https://doi.org/10.1007/s12210-021-01030-z

96   Farnsworth EJ & Ellis DR. 2001. Is purple loosestrife (*Lythrum salicaria*) an invasive threat to freshwater wetlands? Conflicting evidence from several ecological metrics. *Wetlands* 21: 199–209; https://doi.org/10.1672/0277-5212(2001)021[0199:IPLLSA]2.0.CO;2

97   Schooler SS, McEvoy PB & Coombs EM. 2006. Negative per
      capita effects of purple loosestrife and reed canary grass on plant
      diversity of wetland communities. *Diversity and Distributions*
      12(4): 351–363; https://doi.org/
      10.1111/j.1366-9516.2006.00227.x

98   Lui K, Thompson FL & Eckert C. 2005. Causes and
      consequences of extreme variation in reproductive strategy and
      vegetative growth among invasive populations of a clonal
      aquatic plant, *Butomus umbellatus* L. (Butomaceae). *Biological
      Invasions* 7(3); 427–444; https://doi.org/10.1007/
      s10530-004-4063-3

99   Henne DC, Lindgren CJ, Gabor TS, Murkin HR & Roughley
      RE. 2005. An integrated management strategy for the control
      of purple loosestrife *Lythrum salicaria* L. (Lythraceae) in the
      Netley-Libau Marsh, southern Manitoba. *Biological Control*
      32(2): 319–325; https://doi.org/10.1016/j.
      biocontrol.2004.10.012

100   Blossey B, Skinner LC & Taylor J. 2001. Impact and
      management of purple loosestrife (Lythrum salicaria) in North
      America. *Biodiversity and Conservation* 10(10): 1787–1807;
      https://doi.org/10.1023/A:1012065703604

101   Tavernia BG & Reed JM. 2012. The impact of exotic purple
      loosestrife (*Lythrum salicaria*) on wetland bird abundances.
      The American Midland Naturalist 168(2): 352–363; https://
      doi.org/10.1674/0003-0031-168.2.352

102   Radanovic M, Milanovich JR, Barrett K & Crawford JA. 2017.
      Stable isotopes reveal an invasive plant contributes more than
      native sources to anuran larvae diets. *Journal of Freshwater
      Ecology* 32(1): 337–347; https://doi.
      org/10.1080/02705060.2017.1295885

103   Brown CJ, Blossey B, Maerz JC & Joule SJ. 2006. Invasive
      plant and experimental venue affect tadpole performance.
      Biological Invasions 8: 327–338; DOI https://doi.org/10.1007/
      s10530-004-8244-x

104   Maerz JC, Brown CJ, Chapin CT & Blossey B. 2005. Can
      secondary compounds of an invasive plant affect larval
      amphibians? *Functional Ecology*, 19(6): 970–975; https://doi.
      org/10.1111/j.1365-2435.2005.01054.x

105   Kleppel GS & LaBarge E. 2011. Using sheep to control purple
      loosestrife (*Lythrum salicaria*). *Invasive Plant Science and
      Management* 4(1): 50–57; https://doi.org/10.1614/
      IPSM-D-09-00061.1

106   Iancu IM, Bucur LA, Schroeder V, Miresan H, Sebastian M
      Iancu V Badea V. 2021. Phytochemical evaluation and
      cytotoxicity assay of Lythri Herba extracts. *Farmacia* 69(1):
      51–58.; https://doi.org/10.31925/farmacia.2021.1.7

107   Jouravel G, Guénin S, Bernard F-X, Elfakir C, Bernard P &
      Himbert F. 2010. New biological activities of *Lythrum
      salicaria* L.: Effects on keratinocytes, reconstructed epidermis
      and reconstructed skins, Applications in dermo-cosmetic
      sciences. *Cosmetics 4*(4), 52; https://doi.org/10.3390/
      cosmetics4040052

108   Piwowarski JP, Granica S & Kiss AK. 2015. *Lythrum salicaria*
      L. – Underestimated medical plant from European traditional
      medicine. A review. *Journal of Ethnopharmacology* 170: 226–
      250; https://doi.org/10.1016/j.jep.2015.05.017

109   Piwowarski JP & Kiss AK. 2015. Contribution of *C*-glucosidic
      ellagitannins to *Lythrum salicaria* L. influence on pro-
      inflammatory functions of human neutrophils. *Journal of
      Natural Medicines* 69(1): 100–110; https://link.springer.com/
      article/10.1007/s11418-014-0873-5

110   Migliore L, Rotini A, Cerioli NL, Cozzolino S & Fiori M.
      2010. Phytotoxic antibiotic sulfadimethoxine elicits a complex
      hormetic response in the weed *Lythrum salicaria* L. *Dose
      Response* 8(4): 414–427; https://doi.org/10.2203/dose-
      response.09-033.Migliore

111   Treberg MA & Husband BC. 1999. Relationship between the
      abundance of *Lythrum salicaria* (purple loosestrife) and plant

species richness along the Bar River, Canada. *Wetlands* 19(1): 118–125; https://doi.org/10.1007/BF03161740

112 Mahaney WM, Smemo KA & Yavitt JB. 2006. Impacts of Lythrum salicaria invasion on plant community and soil properties in two wetlands in central New York, USA. *Canadian Journal of Botany* 84(3): 477–484; https://doi.org/10.1139/b06-009

113 Houlahan JE & Findlay CS. 2004. Effect of invasive plant species on temperate wetland plant diversity. *Conservation Biology* 18(4): 1132–1138; https://doi.org/10.1111/j.1523-1739.2004.00391.x

114 Tasker SJL, Foggo A & Bilton DT. 2022. Quantifying the ecological impacts of alien aquatic macrophytes: A global meta-analysis of effects on fish, macroinvertebrate and macrophyte assemblages. *Freshwater Biology* 67(11): 1847–1860; https://doi.org/10.1111/fwb.13985

115 Baso NC, Hill MP, Bownes A & Coetzee JA. 2025. Systematic review and meta-analysis of the Enemy Release Hypothesis as applied to aquatic plants. Aquatic Botany 198: 103866 (May 2025); https://doi.org/10.1016/j.aquabot.2025.103866

116 van der Valk AG. 2012. *The Biology of Freshwater Wetlands*. Second Edition. Oxford University Press, Oxford.

About the Author

Kai Aulio is a biologist with a forty-year-long professional career as a research scientist, senior lecturer at the University of Turku, Finland, and as a senior environment editor in a regional daily newspaper. He has published books and over 40 scientific articles in the fields of botany, hydrobiology, and environmental science. Currently, he is an active blogger on various science and cultural topics.